Exc

Special Forces A Detachment, Vie

We would RON (Rest Over Night) in the tree squares; I would my poncho, blow up my air mattress, and hang my body mosquito net under the poncho and over the air mattress. Then I would dig a body trench about a foot deep. Of course, the trench filled with water, but should we receive incoming, I could roll over into it, finding some protection. As daylight gave way to twilight, the ants left the scene to be replaced by mosquitoes. I have never seen anything like it. I could lie on my air mattress and watch the outside of my mosquito net turn darker as the infuriating little insects blocked out the twilight. I would tap the net with my finger and a hole of light would appear, then close up again as the mosquitoes returned.

B-36 FANK Training Command, Vietnam

From December of 1970 to August of 1971, I spent my time training Cambodian soldiers. We received plane loads of C-130s filled with trainees from twelve to sixty-four years old. I often think about those young boys who had to become men all too soon. The Khmer Rouge, under the leadership of dictator Pol Pot, inflicted a nightmarish slaughter between 1975 to 1979, causing the deaths of at least 1.5 million Cambodians. This blood bath surely cost most of the men and boys we trained their lives.

The Korean Special Forces Compound, Seoul, Korea

MAJ Bates, my Army plans officer, handed me several sheets of paper stapled together: "You're not going to believe what I just found in the back drawer of my file cabinet. It's a briefing. No author noted. Not even a classification. If it's true, it has to be Top Secret."

I began flipping through the papers. The first page simply read RAID then MISSION. The next page listed the OBJECTIVES: Hanoi City, Hoa Lo Prison, SonTay Prison, CAS Site #39 Laos, Haiphong Harbor. I continued flipping the pages until I got to ASSETS.

Bates, standing beside me, stabbed the list of four named Americans with his finger. They were credited to have:
-Confirmed information on prisoners
and
-Identified nine out of thirteen prisoner of war camp locations.

Stunned, I sat back in my chair. I recognized all four names, but one flew out at me like a witch on a broom. If I were to believe what I was reading, which I did not, and if it were to get out, it would cause a paradigm shift for all of us Vietnam-era veterans. The name was....

ODA 232 SCUBA Team Fort Devens, MA

My main problem was blisters. As always, the first place that blisters appeared was between my toes. I would pop them with the needle I always carried, but within an hour the blisters would fill back up and hurt like hell.

Years later, a member of the Danish Special Forces would show me a trick to address this problem. But that would be later, and this was now. I rubbed Vaseline between my toes, but that only helped a little bit. Eventually, the blisters started to bleed. This caused concern with infection. Others were having similar problems.

Testing Dry Suits Under Extreme Conditions

After a few swims, we discovered we had to wear ¼-inch neoprene mittens rather than five-finger gloves. When the water hit our faces, it immediately froze. So, for protection, we coated our faces with gobs of Vaseline. This way the ice freezing on our face didn't touch the skin. Of course, our eyebrows caked with ice. Although we wore heavy woolen socks, the cold was so numbing that when we completed a couple of hours swimming with fins, we weren't able to stand, much less walk, for several minutes. Swimming under these severe conditions proved a challenge, to say the least.

ODA 223 Mountain Team Fort Devens, MA

We trained in class one through five climbs. Our training included not only repelling off cliffs but also from helicopters. We trained not only in free climbing, where we always maintained three points of contact with the wall and climbed without a rope, but also in technical climbing, which involved the use of rope, belays, pitons, and metal spikes (usually steel) that we hammered into a crack or seam in the rock. Our mission was not only to become proficient in the art but also to teach basic mountaineering to other Teams as needed.

The Most Fun I Ever Had With My Clothes On

A March from Private to Colonel

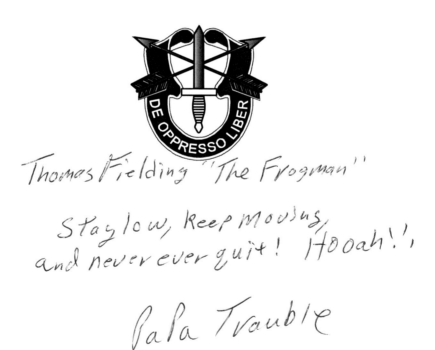

Thomas Fielding "The Frogman"

Stay low, Keep moving, and never ever quit! Hooah!!

Papa Trouble

The Most Fun I Ever Had With My Clothes On

A March from Private to Colonel

COL (Ret.) Tom Davis

Old Mountain Press

Published by:
Old Mountain Press, Inc.
85 John Allman Ln.
Sylva, NC 28779

www.OldMountainPress.com

Copyright © 2014 Tom Davis
Edited by: Sam Zahran
ISBN: 978-1-931575-83-6
Library of Congress Control Number: 2014906612

The Most Fun I Ever Had With My Clothes On: A March from Private to Colonel
All rights reserved. Except for brief excerpts used in reviews, no portion of this work may be reproduced or published without expressed written permission from the author or the author's agent.

First Edition
Printed and bound in the United States of America by Old Mountain Press • www.OldMountainPress.com • 910.476.2542
1 2 3 4 5 6 7 8 9 10

To Polly, one tough Green Beret. HOOAH!

Author's Notes

First and thanks so much for purchasing this book. Second, I don't expect that anyone but my mother, my wife, and my editor, Sam Zahran, will read this epic from cover to cover. And you don't have to. I suggest that you peruse the table of contents and pick and choose what most tickles your fancy. Most, if not all, of the sub-sections within this memoir stand alone.

At the suggestion of some, I've included an extensive list of acronyms at the end of the book. After all, the Army wouldn't be the Army without an ever-changing and voluminous alphabet of organizations, functions, and events. Who could remember them all? Thus the need for a reference guide.

I've tried to be as accurate as one can be when dredging up memories spanning forty years. I'm sure I've not got them all 100% correct, but I think all are close enough for government work. Also, I'm sure I've misspelled more than a few names along the way. I did the best I could here. If, while reading through the pages, you find such an error, please contact me with corrections. If I should run a second printing, which is highly unlikely, I'll make every effort to correct my mistakes.

While I sat writing my memoir, Polly sat beside me writing hers. Often after reading a portion of her work recalling events of the same time and place as those I'd covered, I was left wondering if we were actually there together. I hope those of you who find this book interesting will purchase hers with an eye on comparing the two. Mars and Venus have nothing on us. The prologue to her memoir can be found at the end of this book.

Finally, I tried to compose an acknowledgments page, but the names of those who have meant so much to me along this march from Private to Colonel would have added another ten pages to an already too-long book. However, I think I did a pretty good job of crediting most as I wove my story into the narrative which follows.

Enjoy!

"The only difference between a war story and a fairy tale is that a fairy tale begins with 'Once upon a time...' and a war story begins with 'You ain't gonna believe this shit but...'"

SFC Ronald Brockelman
Team Sergeant
ODA222 and 232 (SCUBA)

Contents

Chapter 1 19
  The First Year 19
    Basic and Advanced Individual Training 19
    OCS and Jump School 24
    The 82nd Airborne Division 27
    The Special Forces Officers' Course 31
    The Vietnamese Language School 34

Chapter 2 49
  Vietnam 49
    Tan Son Nhut Air Base 49
    5th Group headquarters in Nha Trang 51
    ODA 325 Duc Hue 52
    B-36 FANK (Forces Armee National Khmer)
      Training Command 77

Chapter 3 97
  Fort Devens, Massachusetts 97
    Appalachian Trail Walk 100
    ODA 23/223 Mountain Team 105
    Underwater Operations School Key West, Florida 124
    ODA 222 (SCUBA) 129
    ODA 232 (SCUBA) 138
    The Danish Combat Swimmer's School 141
    It's A Small SF World 144
    Special Forces Surface Swimmer Infiltration Course . 145

Chapter 4 159
  Fort Benning, Georgia 159
    The Infantry Officers Advanced Course 159
    Ranger School 160

Chapter 5 165
  Flint Kaserne, Bad Tolz, Germany 165
    ODA 3 (HALO) 165
    Battalion S1 182
    Headquarters and Headquarters Company Commander . 187

Chapter 6 197
   Fort Polk, Louisiana 197
      Combat Support Company 3rd Battalion 10th Infantry ... 197
      5th Division and Installation's Reenlistment Officer 210

Chapter 7 217
   Fort Leavenworth, Kansas 217
      The Command and General Staff College (CGSC) 217

Chapter 8 221
   Fort Bragg, North Carolina–Again 221
      S1/Deputy Commander Support 5th Special
        Forces Group 221
      XO 1/5 SFGA 226
      Deputy G1, 1st SOCOM 229
      Command 3rd Battalion 5th SFGA 245
      Command 2nd Battalion, Training Group 269

Chapter 9 275
   Korea 275
      J3 Special Operations Command Korea (SOCK) 275

Chapter 10 283
   Fort Bragg, North Carolina–Again, Again 283
      Readiness Group Bragg 283
      SLDC Special Forces Command 294
      JSOTF Commander, Incirlik AFB, Turkey 297
      USASOC Inspector General 308
      Retirement Times Two 314

Acronyms 319

Prologue: Stumbling Toward Enlightenment 1
   A Memoir by Polly B. Davis 1

Prologue

"DAD, GOT A call from Ms. Burk." I sat holding the phone in a white-knuckled fist.

"Yeah?"

"Well. . . she said. . . you know. . . it looks like my number's up." I glanced around my apartment in Athens, Georgia. An overstuffed and stained sofa sat against the far wall, framed by two wooden end tables, each topped with lamps sporting large umbrella-shaped shades. I found myself in a probationary summer quarter of law school at the University of Georgia. I hoped against hope that I would do well enough to get accepted as a regular student in the upcoming fall quarter.

Dad's friends nicknamed him "Lightning" when he was growing up because of his v e r y slow drawl. Now he drew out a question to the point that made me want to scream: "What   do   you   think   you'll   do   about   it?"

This wasn't the way I envisioned the conversation going. No way. He was a prominent south Georgia lawyer. My grandfather was a federal court judge for the middle district of Georgia. Both had connections up the political chain all the way to Sam Nunn, senator from the great state of Georgia and the chairman of the Senate Armed Services Committee. For God's sake. I expected him to say something like, "Give me a couple of days to look into this." Then things would be okay, and I would be exempt from the big green Army Machine.

"Well, I don't know. . . Looks like I might. . . You know. . . Have to go in. And, of course, I'll be in Vietnam before long."

"Uh huh," came the distant reply.

In the summer of 1968, things weren't looking very good over there. During that year's Tet Offensive, we lost over 500 Americans, with some 2500 more wounded in just seven days. All together that year we would lose over 16,500 souls. The fact that the enemy was well on its way to losing over 200,000 didn't do much for the hope that I would make it back in one piece, if at all.

I continued, "Well, I looked into the National Guard, but the recruiter told me that they were full up now. Maybe in a month or two an opening would come up. . . and. . . you know. . . I might be able to get in. But I don't have a few months." Okay, for sure. Now I expected to hear him say, "Let me give Sam a call and see what he says." But nooooooooo.

"Looks like you don't have much of a choice then," Dad said instead.

I decided to give him one more chance to pull my fat out of the fire as he had always done in the past. "What do you think I should do? You know... About this."

"I think you should be careful and write home often."

Well, Hell!

YEARS LATER, MOTHER told me that Dad didn't sleep completely though a single night the entire year I spent in Vietnam. He came from that generation, the Greatest Generation, that saw service to your country as a sacred honor and one to be taken on with pride and distinction. I think today that if something happened to my son, daughter, or one of my grandchildren, I truly believe that life would not be worth living. I don't know if I could have done what Dad did. I would hope I could, but I also hope I don't ever have the opportunity to find out.

I DECIDED THAT I would join rather than get drafted. After all, I could get a delayed entry of three months if I joined, and every day I stalled put another day between me and my visit to that tropical paradise. So that's what I did. The Reluctant Warrior climbed into the bus and bounced his way to the induction station in Atlanta, Georgia, for testing and an enlistment physical.

Unloading from the bus at the reception station, I shouldered my way into the large building that would be my home for two days of prodding and probing and test-taking. I had decided that I would apply for Officers' Candidate School (OCS). After all, that could delay my visit to Vietnam another six months!

Two questions on the general knowledge test gave me my first indication that I might actually be embarking on something that I might just make my niche in life: "How do you pour beer into a glass and not form a head?" and "How many balls are there on a pool table?" I kid you not!

AND SO BEGAN, dear reader, my thirty-one year (and, at first, reluctant) march from Private to Colonel–a march filled with joy and tears, happiness and sorrow, success and failure, comfort and hardship. If nothing else, it would prove to be interesting with Polly in lock step by my side. As you march with me through the following pages, my hope is that you'll find it the same.

When my march ended, I could look back and say that, for most part, it was the most fun I ever had with my clothes on.

Chapter 1

The First Year

Fort Dix, New Jersey; Fort Benning, Georgia; Fort Bragg, North Carolina; and Fort Bliss, Texas

Basic Training, Advanced Individual Training, Officer Candidate School, Marrying Polly, Airborne Training, Short Time in the 82nd Airborne, Special Forces Qualification Course, and the Vietnamese Language School

Basic and Advanced Individual Training

I FLEW IN a commercial jet from Atlanta to somewhere in New Jersey, then rode a bus to Fort Dix, New Jersey. I don't remember much, if anything, about the trip, but I do remember very well the final destination. Fort Dix is a sand pit hot as hell in the summer and cold as a witch's tit in the winter. Of all the many locations that the Army saw fit to drop me throughout my thirty+ year career, I have to say that Fort Dix was the armpit.

It seemed that the Louisiana National Guard had filled most of the slots in my basic training class. And a more fun bunch of guys I had hardly ever known. Louisiana Coonasses (their word, not mine), they were there for a good time even in this armpit of a place. Michael Doherty, a tall, round-faced guy with black hair, and Richard (Rich) Anderson, with his sharp nose, V-shaped jaw, and infectious smile always painted on his face, were the two I remember most. We were on the same floor of a two-story wooden World War II barracks. Bunk beds lined both sides of the long room with wooden foot lockers at the foot of each of the bunk beds. I had attended Gordon Military College for my last year of high school, so I had a leg up on all the guys there. I knew well how to spit shine my shoes and make a bed with the tight corners. I also knew the basic Army mentality–it doesn't necessarily have to make sense to do it! This was more than a little hard for my Coonass friends to take in.

I will never forget one guy who was so flat-footed that when he was standing in the shower his ankle bone almost touched the floor. How he made it through the induction physical will forever mystify me. At the end of a day of training, he would hobble around the barracks in obvious pain. Within about two weeks, our platoon sergeant, SFC Cato, a Latino with no detectable accent, had the dispensary reevaluate him. Soon, he left the Army with a medical discharge.

SFC Cato shared a pearl of wisdom with all of us that I never forgot. On the first day, he announced to the entire platoon, while we stood in a somewhat shaky formation, that "the Army took away your rights, and, if it saw fit, would give them back to you as privileges." When I thought about it like that, things seemed to make sense, in some perverse way.

Basic training went pretty much the same for all soldiers—lots of running around in boots, marching to the rifle range through ankle-deep sand, firing all kinds of weapons (often without any hearing protection. Hey, nobody knew), and lying awake in the top of a bunk bed at night listening to all the snoring, coughing, and farting. And, of course, drinking 3.2% alcohol PBRs at 25¢ a can in huge beer halls filled with rowdy drunken trainees. Ahhhhh, could it get any better than this!

The first guard mount I stood went quite well. That afternoon, we had to fall out in formation for inspection. The Officer of the Guard, a second lieutenant, along with his noncommissioned officer (NCO) walked down the line inspecting all of us for our tidiness in dress and spit-shined boots. He also asked questions of each of us, testing us on what all we had retained from the classes that most of us dozed through.

As previously mentioned, I had had some experience with Highschool ROTC at Gordon. Here we drilled with the M1 Grand rifle. I remember the Officer of the Day asking me to list the type of M1 ammunition. I rattled it off and ended with the blank that was used to propel a grenade from the M1. Apparently, he had never heard of this one and turned to his NCO with a question on his face. The old NCO smiled and nodded.

That did it! I was selected as the Colonel's Orderly, and as a result did not have to stand guard duty that night. Instead I spent the day outside the Brigade commander's office doing mostly nothing. I did have an interview with the Colonel. He told me that he was writing a letter of commendation for me and including it in my file, and he'd be sending a copy to my family. Dad thought it was pretty cool when he received it.

Meals were a welcome occasion in basic training, and your place in line was of great importance: the closer you were to the front, the sooner you got to eat. I remember one day Mike, Rich, and I were standing in line, and these two guys from their unit decided that they would break up in front of us. This was something you just didn't do, and I brought it to their attention.

Well, this didn't go over too well with them. They got into my face, asking what I thought I was going to do about it. I had taken a year of Karate at the University of Georgia from a 10th degree black belt from Thailand, Per Son Wongui (not sure about the spelling here, but that is what it sounded like). I worked my way up to a brown belt and, in the process, fought two and sometimes three opponents.

Both of these guys were as big if not bigger than I was, but I was pissed, to say the least. I looked at them and said something to this effect: "Okay. You guys want to make something of this. I'll be happy to kick your ass all over the area, but I want both of you at the same time. After all, when the shit hits the fan and SFC Cato wants to jack us around, I want to be sure that he knew I didn't take advantage of you." Well, things went pretty quiet in the chow line, and the two guys looked at each other, then back at me, then fell in at the back of the line.

Rich Anderson nudged me and said, "I know those two. What the hell were you thinking? They could've kicked your ass all over the place."

I smiled and said, "Yeah, but, apparently, they didn't believe that."

COMPETITION AMONG THE companies in basic training grew fierce. We were graded on the average scores on the various written tests as well as physical training and marksmanship. As it turned out, most of the guys in my company (the National Guard ones) had college degrees, so we were way ahead in the academic area. An incentive to score high in marksmanship was a three-day pass for all who fired expert with the M14. As luck would have it, I did, but neither Mike nor Rich did. They were devastated.

I and another guy I liked named Dewitt (don't remember his first name) from Illinois decided to go to New York City on our pass. Dewitt's shoulders were wide as an ax handle, just the type of guy you wanted next to you if you got in trouble with a bunch of hippies wanting to give you a hard time.

A quick small-world story: Dewitt and I, with shaved heads and wearing our army dress greens, were leaving the play *Man of La Mancha* when to my rear I heard someone call, "Tom. Tom Davis. Hey, is that you?"

I turned to see a girl whose face was vaguely familiar, but I couldn't place her. She had a friend with her. They were closing in on Dewitt and me.

"Don't you remember me?" She told me her name and then it came to me. "I was in Mexico with you in summer school!"

After failing Spanish 101 twice at Georgia Southwestern, Dad sent me down to Monterrey, Mexico, for a crash course in Spanish. I managed to not get any credit for Spanish down there either, but that is another story that my buddy for that fiasco, Tommy Tucker (now of Tucker's Bar-B-Que in Macon, Georgia), will have to include in his memoir.

I couldn't believe it. Here I was in New York City and had run into someone I knew down in Mexico. Dewitt did his best to convince the two to have drinks with us, but to no avail. I wasn't too keen on that as I was pining away for the girl I'd dated throughout my Junior and Senior year at UGA, Polly Brown of Macon, Georgia, fame. Polly, a tall willowy blonde, a little flat chested, legs to die for, always flashing a gorgeous smile, was now my fiancé.

AT THIS POINT I have to digress. I've made some really bad decisions in my time and some really good ones. Before entering the Army, I met Polly at the Atlanta airport. She had just returned from her trip to Europe. That night in the car heading south on I75 to Macon, Georgia, I turned to Polly and said, "I think we oughta get married."

She didn't say anything at first, pretending she didn't hear me.

"Well, what do you think?" I said, keeping my eye on the highway.

Polly turned and looked at me. "Well, I had thought when you asked it would be a little more romantic, but I guess that would be a good idea." I wasn't sure, but I thought she might have smiled while saying it. When we got to Macon, I asked Pat, Polly's dad, if he thought it would be okay if I married his daughter. He allowed how it would be.

I have to tell you, dear reader, this was the best decision that I have ever made before or since. Not only was and is Polly an incredibly beautiful person, she is one of the most strong-willed and determined individuals I have ever known. She would weather thirty years with me in the Army, become the mother to two great kids, and do it in grand style all the while retaining a sense of humor. Now, I'm not writing this to get brownie points with my wife. I truly do mean it although, at times, I know I might not show it. My bad.

TIME MARCHED ON and Christmas of 1968 produced a break between Basic Training and Advanced Individual Training (AIT). The Army gave us two weeks off. I flew home for Christmas. Polly Brown picked me up at the Atlanta airport, and we spent the time bouncing back and forth between Vienna and Macon, something we would do constantly for the next thirty years.

CHRISTMAS LEAVE ENDED and found me back at Fort Dix in a foot of snow, a private making $78.00 a month, and wondering how in the world I was going to support a wife. I was scheduled to go to Infantry OCS at Fort Benning, Georgia, but the graduation rate at that time was less than 50%. I told Polly that if I didn't make it, we would have to postpone the upcoming wedding until I got back from Vietnam, assuming I got back from Vietnam in an upright position.

WHEN I GOT back to Fort Dix, I was slated to go to a training company comprised mostly of those of us who had been accepted to attend OCS. As a result, I found myself separated from my Coonass buddies. The barracks we stayed in were three-story and cinder block. The large open bays and communal bathrooms were basically still the same though.

Advanced Individual Training (Infantry) placed emphasis on squad, platoon, and company tactics, the different weapon systems in the company, M79 grenade launcher, 106mm recoilless rifle (an anti tank weapon), the 60mm machine gun, marching, and so on. We marched to and from classes and to the ranges in deep sand chanting, "Dress it right and cover down, forty inches all around, that's the Fort Dix Boogie, what a crazy sound." Of course you could substitute any other army training base for Dix and you had the universal chant.

Now, speaking of the 60mm machine gun, I remember being on the firing range when a deer happened to cross the open area to our front. Among the cries of "Cease Firing!" trails of red tracers streaked across the range painting the ground all around the hapless animal. Zigzagging through the various pop-up targets, the deer finally made it across and into the woods. Not a scratch on him. Oh, bad to the bone we were. Really was feeling sorry for what we'd do to Charley when we all got to Nam.

The one guard mount I stood in AIT turned out to be a rerun of the one in basic training. Believe it or not, I was again selected as the Colonel's Orderly and received another letter of commendation. I remember discussing with the training brigade commander my attending OCS. He thought it was the right thing to do. Oh, well. . ..

OCS and Jump School

I'D DECIDED THAT I would go to OCS for several reasons. Money—I would be getting paid as an E5 (a buck sergeant's pay) while there and as a 2nd Lieutenant should I happen to graduate. Also, it would be management experience, something I thought might enhance my basically blank resume. And, oh by the way, it would delay my departure for Vietnam by six more months!

What can I say about OCS. Basically, it sucked. Well that may be an overstatement. It seemed to me that the objective of all in charge was to do everything that could be done to force those new OCS candidates (who were referred to as Blue because of the blue helmets we had to wear) to sign quit letters. Something about making us mentally tough. Right.

Having said that, I must add that the training was pretty good at the small unit leader level. The things I remember about OCS besides the harassment (only about 50% of those who started finished) was running everywhere we went in formation. We stood in line at the chow hall only to get food on our trays, then eat it as fast as we could as we walked toward the trash cans, depositing any left on our plate into the garbage as we continued out the door headed to our next class. It seemed that we were always in a hurry to get there, then stood around, waiting for things to get started. Go figure.

My first month in OCS, I made the unforgivable mistake of leaving the lock on my footlocker unlocked. Never mind that the footlocker had to remain open for inspection at all times. See what I mean about sucking? Anyway, the platoon's tactical officer, a 2nd Lieutenant Chaffin, who had himself confused with God, called me down to his office and began to chew my ass. During the course of the "counseling session," he told me that he was going to make it his mission in life to run me out of OCS, and I might just want to sign a "quit letter" now. He then pushed the paper toward me and smiled.

Okay. About the only way to get me to do something that I wasn't really sure I wanted to do was to tell me I couldn't do it. I'd had just about all the crap I could stand from this guy anyway, so I looked at him and told him, "You may be able to flunk me out of this program, but I guarantee that you will never under any circumstances ever get me to sign a damn quit letter!"

I actually thought the guy was going to have a stroke right there in front of me. Instead he jumped up and stuck his crimson face in mine and said, "Well we'll just see about that!" and ordered me to grab hold of the pipes running across the ceiling and hang there until he told me to let go.

I figured I'd pushed it about as far as I should, so I did what he ordered. I'd hold on until I couldn't any more, then drop to the floor where upon he would scream at me to get back up and hold on longer. This went on until I just couldn't do it any more. He again shoved the paper in my face. I shook my head and told him, "No damn way!"

Finally, he gave up and told me that this wasn't over. And it wasn't. He stayed in my stuff for the next several months. Once again, I'd let my alligator mouth overload my hummingbird ass. This was a prelude of things to come over the next 30 years!

All the guys knew what had happened and rallied around me. No matter how bad a tac officer wanted to throw someone out, if the candidate had good peer reports, the officer was fighting an uphill battle. In my case, even the guys who might not think I was such a good guy would say good things about me just to stick it to the tac officer.

We had a system of peer review in OCS. Once every two weeks or so, we had to write what were called Bayonet Sheets on, I think, about five guys. Either telling what they did that was good or what they did that showed they weren't officer material. We had to write a bad report on at least one candidate. As a result of this requirement, any candidate who even hinted that he was thinking about dropping out would get a bad report from others. The point being that if you had to write up a bad report, you wanted it to be about someone who didn't really want to be there in the first place.

By the time we turned "Black" (meaning we wore black helmets and were within a couple of months of graduation), LT Chaffin had gotten orders for Vietnam, and a really good guy took his place. Life for me in OCS became bearable. I don't know for sure, but I heard that LT Chaffin had gotten himself fragged. Probably didn't happen, but I could see how it might have.

As graduation drew near, we put in our "Wish List," a list of assignments we would like to have. Not that the personnel assignment people cared, but the regulation required them to ask. Of course, the top five in the class got their first choice. Needless to say, I wasn't one. I had volunteered to go Airborne and to Special Forces (SF). I figured that if I was going to go to Vietnam, I wanted to have the best training possible and be with the best guys in the Army, ergo, Special Forces. And besides, attending Airborne training and the Special Forces Qualification Course (SFQC) would delay my visit to Vietnam by at least four or five more months! And there was the added benefit of jump pay–$110.00 a month, which would bring my lieutenant's pay up to around $450.00 a month!

The word came down that if I wanted to go to Special Forces that I would have to go Voluntary Indefinite (VI). This meant I would effectively add a one-year extension to the two-year obligation I incurred when I graduated from OCS. No way. Two years was all I had signed up for and was all I was going to give this man's Army.

In the end I came down on orders for Jump School, staying at Fort Benning for three more weeks, then on to the 82nd Airborne at Fort Bragg, North Carolina. Well Hell.

While in OCS, I was able to see Polly only on rare occasions. I still maintained that unless I graduated (something that often seemed like it might not happen–if Chaffin had anything to say about it), we would have to put the marriage off until and if I made it back from Vietnam.

As it turned out, I did graduate, and we were married in the garden of her grandparents' home in Baconton, Georgia. I stood in my new dress blue uniform surrounded by angry gnats and dripping sweat on a hot-as-hell July afternoon in South Georgia.

We honeymooned that weekend in Panama City, Florida, and I started Jump School the next Monday. For some reason, I always have attended these macho Army schools at the wrong time of year– Airborne in the summer in Georgia, SFQC in the winter at Fort Bragg, Underwater Operations in Key West, Florida, in the winter (the winter they had snow in Orlando, Florida.), Ranger school in the mountains of North Georgia in the winter, and so on.

I found a nice little one-bedroom apartment that we could afford just outside Fort Benning in a somewhat less than desirable location. Hey, it had a bed, an indoor toilet, a refrigerator, and a stove. What more did we need? Right? We moved in and set up living for the short three weeks that was Jump School.

Polly was determined to iron my fatigues every day while I attended Jump School. Well that lasted about two days. She says three, but it really was two. Anyway, after three weeks of running around in the August heat in boots and jumping five times out of the old C-119 cargo planes, I was awarded my airborne wings, and Polly pinned them on me. Then we packed up and moved to Fort Bragg, North Carolina. Hooah!!

The 82nd Airborne Division

IN AUGUST OF 1969, Polly and I pulled into our newly rented house just off Murchison Road (AKA The Murk). A friend who graduated with me from OCS, Dewy Newman, arrived a month before we did and found the place. After we moved in, Polly found out that this was not "the" place to live. Hey, it had a bed, indoor toilet, refrigerator, and stove. What more did we need? Right? And when you opened the cabinet doors under the sink, you didn't see the ground (that story is saved for later in the book).

By the end of that weekend, we had settled in to our little Shangri-La, and I reported into the 2nd Battalion of the 325th Infantry in the 82nd Airborne Division on Monday. I should have known something was up when I passed the command structure board hanging in the hallway with all the photos of the commanders and platoon leaders on it, each listing their source of commission. The battalion commander and all of the company commanders and line platoon leaders were West Pointers. Coincidence? Hardly. But that went right over my head.

The battalion's Command Sergeant Major (CSM) directed me to knock on the Battalion Commander's door and present myself. LTC Prynne, The "Old Man," was expecting me.

The interview went something like this: "Sir, Lieutenant Davis reporting for Duty." Snappy salute.

"Very good, lieutenant. Have a seat." LTC Prynne waved toward a chair sitting just in front of his desk.

I sat, and we made small talk, mostly about my background and what would be expected of me in the battalion. It was at this point that LCT Prynne asked, "Well, what do you think you'd like to do?"

Of course he meant what would I like to do within his battalion. I, of course, took that to mean what did I want to do at Fort Bragg. Hey,

I wanted to go to Special Forces, but I didn't want to give Uncle Sam a third year.

OKAY. IT'S AT this point that all you readers who are my date of rank or senior are beginning to smile. 'Cause you know what's coming.

"Well, sir, I volunteered to go Airborne and Special Forces out of OCS, but they only gave me Airborne and sent me here." I looked around the room. "What I'd really like to do is go over to the Special Forces."

Hey, it was an honest answer to his question. Right? As the red crept up the man's neck toward his face, he shifted in his chair, picked up a pen, and tapped it on his desk while filling his cheeks with air. I could see by the blowfish look he was sporting that things were going a little sideways here, but I didn't understand why. You would've thought I'd stood up, done an about face, dropped trout, and mooned him. And, in effect, that's just about what I had done. Although I didn't know it at the time.

"Well, Lieutenant (emphasis on the *Lieu*), why don't you just give us a chance here in the 82nd before you decide to go over there!" He made the words "over there" sound like "pedophile" and jabbed his finger in the general direction of where the U.S. Army John F. Kennedy Special Warfare Center and School (USAJFKSWCS)–known informally as SWC–sat on the other side of Fort Bragg.

I LEFT THE Battalion headquarters and walked across the street to Company B. When I got there, the First Sergeant told me to go directly into CPT Harding's office. He was waiting. I basically went through the same routine of formally reporting into Harding that I had with Prynne. When I finished, Harding, a tall big-boned man with a long, acne-scared face and intelligent eyes, commenced to chew my ass out about the negative attitude I had "displayed" a few minutes ago with the Battalion Commander. The fact that Prynne had asked me a question, and I gave him an honest answer didn't seem to matter.

Strike one for the 82nd.

An extra year to get outta here and into SF was looking more doable by the minute.

The interview with Harding ended, not nearly soon enough for my taste, and I left to meet my platoon sergeant, SSG Beddingfield of the weapons platoon. SSG Beddingfield was a round soldier. I mean that he was literally round (the Army hadn't gone all postal about tubby troopers at that time), with a big, welcoming smile on his face. He was a veteran of both Korea and Vietnam and had forgotten more about indirect fire weapons than I would ever know or care to know. He was a very good soldier.

After an introduction to the members of the platoon and a tour of the motor pool and the barracks, SSG Beddingfield told me that the platoon would be supporting the STANO (Surveillance Target Acquisition and Night Observation) testing with illumination starting the next night for the next seven or eight weeks. Giddy up!

We packed up every Monday and unpacked every Thursday, using Friday to clean equipment for the next several weeks. Since I wasn't available to pull duty officer during the week, the company commander told me that I would pull it every other weekend.

Strike two for the 82nd.

As any gung-ho, brand spanking new 2nd LT assigned to an Airborne unit would, I joined the 82nd Sports Parachute Club. When I wasn't in the field or pulling duty officer, I would drag Polly out to the Raeford drop zone (DZ) so she could watch me be the MAN. Of course she had to sit around for two hours or so just to watch me bask in my ten minutes of fame.

She soon got bored with the whole thing, and while I was out doing my thing at night during the week, she signed up for classes with the parachute club and announced that I could now go out to the DZ with *her* and watch *her* jump out of airplanes. Well, Hell!

On Polly's very first jump, she went straight for a grove of pine trees and proceeded to destroy several as she tumbled ass over tea kettle through them. I concluded that it was a hell-of-a-lot easier jumping out of the plane than watching someone you loved do it. My fun meter was not registering so high. I suggested that she hang it up and just be happy watching me be a MAN. That didn't go over too well either, and the next Saturday she was right back out there defying death and turning me into a nervous wreak. As it happened on her second jump, she managed to land near the pit! Hey, even a blind hog finds an acorn now and again.

Much to my pleasure, she declared that she had mastered the art of parachuting and was going to give it up. Did I mind? Not at all.

SPEAKING OF JUMPING out of airplanes, specifically jumping with the 82nd Airborne, if you think that being busted for drugs in Turkey you're in for the hassle of your life, then you haven't jumped with the 82nd. I made two mass tactical jumps while with them. You had to arrive three or four hours prior to the jump. Then go through a list of pre-jump training a mile long. Next load into a C-141 with about 100 other jumpers. Then fly low level so the Air Force guys could get their nap-of-the-earth training in. And sit sandwiched in between two guys throwing up their entire breakfast and lunch. And, finally, getting the "STAND IN THE DOOR" command, you put your knees in the breeze, stepping out into the wild blue yonder with your 50-pound rucksack hanging underneath your reserve. If the chute opened, that was a good thing, but getting out the door was great! Compare this to jumping with Special Forces which took about one tenth the time with way less hassle.

Now, I have to stand up a little for the 82nd here and say that the average age of the 82nd jumper hovered around twenty and the average number of jumps about ten. The average age for the SF jumper was thirty-two and the average number of jumps about thirty-five. Add to this that the 82nd was dealing with well over 150 jumpers in several plane loads while the SF guys might put thirty or so out the door. You can see why the 82nd was the 82nd Airborne.

ABOUT THREE MONTHS into my time with the 82nd, I got called into the company commander's office. He presented me with a written counseling statement. Apparently, the company had received a call from the MP's notifying them that I had left my car unlocked and that a ticket was coming down through the chain of command. Lord help us!!

Strike three for the 82nd.

I signed the counseling statement, acknowledging my horrendously bad behavior, walked over to the battalion headquarters, and called Infantry branch at the Military Personnel Center (MILPERCEN) in Washington, D.C.. I told an assignment officer there that I was just dying to give the Army one more year and would they pleeeease send me to the Special Forces Qualification Course (SFQC or Q Course) and maybe even to language school after that. Hey, language school would take at least three months to complete, and that would mean three more months before I had to do my duty to God and Country in Vietnam.

I LEFT THE 82nd, Officer Efficiency Report (OER) in hand, just before Christmas of 1969 and didn't let the door hit me in the ass. After all, the door would just have shoved the OER further up there than it already was. And we couldn't have that. Could we? *Auf Wiedersehen!*

The Special Forces Officers' Course

THERE BEING A great need for SF qualified officers, the school was cranking us out hand over fist in the three-month long course. In the song *The Ballad of the Green Beret*, one stanza goes: "One hundred men we'll test today. Only three will win the Green Beret." Being the irreverent students we were, we changed that stanza to, "One hundred men we'll test today. A hundred and one will win the Green Beret."

I joke about this, but to tell the truth, having eventually attended civilian schooling through a Masters Degree and having lost count of the number of military schools I've attended up to and including the Army War College, I can say without a doubt that the Q Course was the best training and instruction I had received to that point or have received ever since. Bar none.

Early in the course we had a guest speaker whose name I can not recall. He was an old, weathered retired colonel and had jumped into France with an OSS Jedburgh Team during World War II. He began with a hard landing, knocking himself out. When he came to, he was staring into the face of the most beautiful woman he had ever seen. A member of the French Resistance and a woman who would become his wife. And still was.

I sat spellbound for the two-hour lecture. At the end of the lecture, I knew that I had found my destiny. It was clear to me that no matter what happened, I wanted to–no, I had to–become a Special Forces Soldier. And I did.

THE SF OFFICERS' course was heavy on the academics. I remember being extremely impressed with how well all the instructors presented their classes. All were combat veterans with several tours in Vietnam. Classes ranged from map reading and survival, to the art of Unconventional Warfare (UW), to small unit tactics of the raid, ambush, reconnaissance, and so on. The instructors also covered all aspects of the specialities found on the Operational Detachment A (ODA), the building block of SF. These included medical, communications, light and heavy weapons, operations and intelligence, and demolitions. We received extensive training in all types of airborne operations, from setting up a Drop Zone (DZ) to setting up a Landing Zone (LZ) for both fixed wing aircraft and helicopters, and calling in MEDEVACs and air strikes.

Probably the most interesting block of instruction was survival training. I particularly remember having been told that a field-expedient means for cleaning a wound was to take a leak and let the first part out then hose down the wound with the rest. As it turns out, urine is basically sterile water. Another tidbit was a field expedient treatment for diarrhea. The instructor told us to heat up fresh killed animal's bones until they became brittle, grind them into a power, mix with water, and drink. Diarrhea was something I was familiar with, and over the thirty years, I would be able to proudly claim that I had had it in all four corners of the earth.

The course had within it two field training operations: one consisted of raids and ambushes or Direct Action (DA) missions and the other, Unconventional Warfare. Our class was broken down into twelve-man sections that mirrored a Special Forces A Team. When we went to the field, we rotated the Team leader position, and each of us acted as one of the specialities found on the Team. As an example, during the big UW exercise, Gobbler Woods, when I was not the acting Team leader, I played the role of the Team's engineer. As the engineer, I gave demolitions classes to the guerrillas, or Gs as we called them. The Gs were privates and noncommissioned officers from the 82nd.

At the time, it was unheard of for any faculty advisor (an officer assigned to each of the student Teams) to flunk anyone during a field operation. As luck would have it, my Team's advisor/grader was a lieutenant colonel (LTC) named Bellfy. He was a former commander of Command and Control North (CCN), one of three command and control units SF had in Vietnam. They conducted strategic reconnaissance missions in the Republic of Vietnam (South Vietnam), the Democratic Republic of Vietnam (North Vietnam), Laos, and Cambodia; they carried out the capture of enemy prisoners, rescued downed pilots, and conducted rescue operations to retrieve prisoners of war throughout Southeast Asia, and they conducted clandestine agent team activities and psychological operations. The short version is that these guys were very, very, very bad asses.

Anyway, Bellfy took the field training and the evaluation portion of the Q Course very, very seriously. In fact, for the first field training exercise (raids and ambushes), he failed every one of us!! Well I've got to tell you that got my attention. The realization that I might just not pass and be awarded a Green Beret suddenly became a reality. Needless to say we were all nervous going into the final UW field training exercise. True to his nature, Bellfy failed two of us and awarded major deficiencies to all the rest except me. He gave me a very good evaluation. Why? To this day, I can't tell you. Go figure.

WE HAD RETURNED from our field training exercise. Classroom training now filled my time. One evening just after supper, I stood in our little livingroom/kitchen when the phone rang. I stepped over to the wall where it hung and lifted the receiver, "Lieutenant Davis."

Dad spoke in his painfully slow drawl from the other end. "Ruby just got a call."

Ruby Joiner was the mother of Johnny, my best friend since Aunt Jennie's kindergarten. Ruby was my second mother, and Johnny's three brothers, Clark, Mike, and Pat, along with their sister, Connie, were like my own brothers and sister. I'd been held back in the fifth grade, so Johnny was one year ahead of me graduating from college and entering the Army. He currently served as an Infantry Second Lieutenant in Vietnam.

Don't say it. Don't say it. Don't say it. I thought as I stared at the grocery list that Polly had pinned to the cork board over the phone.

"Johnny's   been," Dad went on.

Don't say it. Don't say it. Please don't say it. I thought. I could feel my stomach tighten and getting ready to heave up the last of the fried chicken and mashed potatoes that Polly had picked up for our supper.

"Hurt. But he's going to be okay."

Why couldn't Dad have started with that. The relief I felt cannot be expressed. We went on to discuss what all was going on with me, and then I hung up. Several years later when I'd be a Colonel stationed back at Bragg, I'd get a similar call from Dad about Johnny, but it wouldn't be good news.

ON 27 MARCH 1970 I marched across the stage at the JFK Center and was awarded my Green Beret by the commandant, COL Edward M. Flanagan, Jr.. As always, Polly stood there with a big smile and a loud clap. Years later, I would retire from the Army on that very same stage.

With the butter bar of a second lieutenant pinned on my Beret, I marched off the stage and headed to Fort Bliss, Texas, to attend the Vietnamese Language School for three LONG, LONG, horribly LONG months.

The Vietnamese Language School

WE LEFT FORT BRAGG and headed west, stopping in New Orleans, Louisiana, where we linked up with my old basic training buddy, Michael Doherty, and his wife, Danny. Needless to say, The Big Easy was one hell of a town for this country boy and his young wife in 1970! We ate pompano in the bag at the Court of Two Sisters and drank hurricanes at Pat O'Brians. We went to the My Oh My club out on Lake Pontchartrain and ogled female impersonators. And listened to Fats Domino pound the keys and sing at a club on Bourbon Street. Good times were had by all! And that was a good thing, given the next three months I would spend in my own personal hell–the Vietnamese Language School. Ugh!

HOW CAN I describe our tour at Fort Bliss in the Vietnamese Language School? *It was the best of times. It was the worst of times.*

IT WAS THE best of times because of the location. From June to August of 1970, Polly had nothing to do but cook and lie around the apartment's swimming pool. I had nothing to do but eat, lie around the pool, and go to school. We also took in all the sights: White Sands National Monument, Carlsbad Caverns National Park, the Gila National Forest, Elephant Butte Reservoir, and of course *Ciudad Juárez*, Mexico, to attend the dog races, and Ruidoso Downs to attend the horse races. Mother and Dad came out and brought my eleven-year-old brother John with them. When they left, they also left John with us. As I would do numerous times in the future, I exposed him to the finer things in life. We went to see matadors kill bulls. We also went to the dog races where John bet the equivalent of $1.00 in pesos on a dog named (since John was from Georgia) Georgia Hop and quadrupled his money! I'm not sure, but I think I let him sip from my *cerveza* a time or two.

IT WAS THE worst of times because I had to go to language classes every day and listen to tapes in the evening, trying in vain to learn enough Vietnamese to get a basic rating of 1/1 (the first number being the rating for reading and the second being the number for speaking). Often at the end of the three-month course, the students would get a 1+/1+ or even higher rating. Also, language school was the first time I had been off jump status (read that not getting my jump pay). The loss of the $110.00 monthly put my pay at around $340.00 a month.

I had a bad track record when it came to learning a foreign language. I flunked French in high school. In college, I failed Spanish 101 twice, 102 once, 103 once, and passed 104 the first time. Hey, even a blind hog finds an acorn now and again. After I failed Spanish 101 at Georgia Southwestern, Dad sent me down to summer school in Monterrey Mexico for the summer between my freshman and sophomore years. I was hopelessly lost, so a guy named Tommy Tucker (now of Tucker's Bar-B-Que in Macon, Georgia) and I left school for a week and went to Mexico City to party. All went well until the return trip to Monterrey where I spent the night on the toilet in our train's two-person sleeper compartment. It was bad for me, but even worse for Tommy.

It was also a strain for my language instructor, a pretty little Vietnamese lady who was VERY worried that an officer attending her class was actually not going to get a passing grade. Apparently, this had never happened, and she was worried how it would reflect on her. I assured her that it wasn't anything she did or didn't do. In the end I received an academic OER stating that I had attended the three month course, tried very hard to pass, but was leaving with a 0/0 rating. Well, life would go on. And we were soon heading west to Travis Air Force Base (AFB) then on to Vietnam.

En route to Travis Air Force Base (AFB)

THE LANGUAGE COURSE ended in the beginning of August 1970, and I was promoted to 1st Lieutenant. Although the thought of leaving for a year's tour in Vietnam loomed over our heads like a thunder cloud, we did enjoy the trip from Bliss to Travis AFB, California. My trip from Travis to Vietnam and especially Polly's trip from Travis back to Georgia were another thing entirely.

WE HIT ALL the sights on our trek west. Las Vegas was our first stop. We stayed at the Stardust. It was expensive, so we stayed only two nights. I'd make up for it with our accommodations at the Grand Canyon.

I had been sleeping under a poncho hooch for the better part of a year, so when we made it to the Grand Canyon's camping area, I pulled out two large ponchos, hooked them together, strung them up, blew up two US Army air mattresses, and declared to Polly, "Our room is ready!"

As usual, she took it all in stride as she would for the next 29 years. We settled in for the night. In the morning when I woke up, Polly was not beside me. I looked around and found that she had slid out from under the ponchos and was sound asleep in the clear morning air.

We had talked about riding the mules or donkeys (I don't remember which) down into the canyon. Now remember that all Polly had done for the last three months was cook, and I. . ... When we got to the tour office, we found out that there was a weight limit of 200 pounds per person. I looked at Polly. She looked off to her right, then down.

"We'd like two tickets for the tour into the canyon, please." I said to the lady dressed in the green park service shirt and khaki pants who sat behind the counter.

She leaned forward to look me over and asked, "Sir, how much do you weigh?"

"Oh, about 190 pounds. . . I think. . .." I gave her my best South Georgia smile.

She reached under the counter and slapped a set of scales down and said, "Let's see."

I put the scales on the floor and stood on them. The needle hit 210. I had gained almost twenty-five pounds during my eat-fest at Bliss.

We drove in silence back to the camp site, rolled up the ponchos, deflated the air mattresses, got in the old green Chevy Nova, and continued our trek west.

WE MADE IT to Los Angles, California, and stayed with Linda Odom, one of Polly's childhood friends who was doing her residency at Mount Sinai as a psychologist. While there, we went to Anaheim and visited Disney Land. We also visited Universal City and walked through the University of Southern California's campus. Years later, while assigned to Germany, I would get my masters from there.

WE LEFT LA and traveled up Highway 1 through Big Sur, stopped for a wine tasting, visited Hearst Castle, and then on to San Francisco. We had made arrangements to stay in San Francisco with Mike Sullivan, one of the guys on my Team in the Q Course. His dad was a vice president of Bank of America there. Mike was on a thirty day leave en route to Vietnam, too.

We saw the Golden Gate Bridge and ate at Fisherman's Wharf. It was here that my sister Marrlee, my first cousin Julia, and Polly's brother

Malcolm met us. All teenagers. They had flown out so that they could keep Polly "company" on her trip back to Georgia. Right.

Travis AFB sat close to San Francisco. We all got hotel rooms where we stayed until I left for Vietnam. It was now close to the end of August. I don't remember how the final goodby went. I do remember having to share that moment with the whole crew.

WHAT I DO remember about the long, long, long seventeen hour flight to Vietnam is being waked up, it seemed like every three hours, and fed steak. But as bad as that flight was, it was like *Little Ned's Third Grade Reader* compared to Polly's trip back across country with the teenagers. After hearing her accounts of the journey (which she will cover in her memoir), I decided that my trip to and year long stay in Vietnam was just a lovely walk through the park.

Polly would attend graduate school at the University of Georgia while I attended graduate school at the University of Vietnam. She needed more money than I, so when I got in country, I had my pay set up so that $100.0 would go into my Soldier Saving Account, $50.00 would come to me, and the rest to Polly.

Polly pinned on my airborne wings, and we packed up and moved to Fort Bragg, North Carolina. Hooah!!

Polly signed up for classes with the 82nd Parachute Club and announced that I could now go out to the DZ with her and watch her jump out of airplanes.

If you think that being busted for drugs in Turkey you are in for the hassle of your life, then you haven't jumped with the 82nd.

The Special Forces Officers' course had within it two field training operations, one consisted of raids and ambushes or Direct Action (DA) missions and the other Unconventional Warfare.

In New Orleans, LA where we linked up with my old basic training buddy, Michael Doherty and his wife, Danny.

And listened to Fats Domino pound the keys and sing at a club on Bourbon Street.

We went to the My Oh My club out on Lake Pontchartrain and ogled female impersonators.

Mom, Dad, John, and me at the Gila Cliff Dwellings. Polly took the photo.

Polly had nothing to do but cook and lie around in the apartment's swimming pool. I had nothing to do but eat, lie around the pool, and go to school.

In the Grand Canyon, I pulled out two large ponchos, hooked them together, strung them up, blew up two US Army air mattresses, and declared to Polly, "Our room is ready!"

While we were in LA, we went to Anaheim and visited Disney Land.

We left LA and traveled up Highway 1 through Big Sur, stopping for a wine tasting, visited Hearst Castle, and on to San Francisco.

Left to Right Marrlee, Me, Julia, Polly, and Malcolm.

We ate at Fisherman's Wharf.

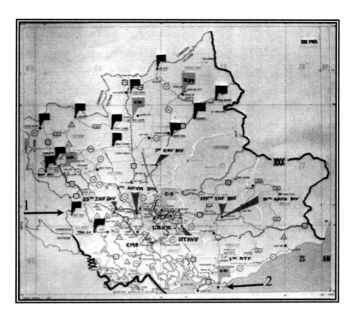

Chapter 2

Vietnam

August 1970 to August 1971

Tan Son Nhut Air Base

OUR PLANE LANDED at Tan Son Nhut Air Base near Saigon, the capital of South Vietnam. When we stepped off the plane, we were immediately aware of the humidity that blanketed us. Then, the smell. Anyone who went to Vietnam still remembers well the smell of burning feces. There was no shortage of 55 gallon drums. They were cut in half and placed under the holes in the outhouses that sprang up like mushrooms around the country. When nearly filled, the half drums would be dragged out, topped off with diesel fuel, and set on fire. Once all the feces had burned out, the drums were placed back under the holes in the outhouse. Actually, a very efficient process.

Buses hauled us to the reception center. We off loaded what little gear we had and moved into the transient barracks where we would stay until our in-country orders came down. These barracks looked similar to the open bay living I had experienced in basic training. The noticeable difference was that all the cots were tented with mosquito netting.

At night, I was treated to the same choirs of snoring, coughing, and farting; only now in the background mosquitoes whined.

After unpacking, we headed to the All Ranks Club and proceeded to drink beer. I was feeling pretty pleased with myself as I had orders for the 5th Special Forces Group in hand.

In stormed a guy who had arrived in country just a couple of days before me with orders to the Group. I had met him at the reception center earlier in the day.

He climbed on the bar stool next to me and said, "I can't believe this shit!"

"What?" I said and turned up my Pabst Blue Ribbon.

"I'm so screwed! They're sending me to the First Cav! Crap!" He ordered a Black Label, squeezed the red and black can until it bent inward, and shook his head.

"I thought you had orders to Group." I sat my PBR back on the bar.

"I did, but this wimpy major in the officer assignments section," he nodded in the direction across the compound where the personnel building was located, "gave me that 'needs of the Army' bullshit."

I sat there, staring at my beer and thinking; then I stood up and walked out the door straight for the personnel building. When I got there, I went to the officers' assignment section and found the noncommissioned officer in charge. I hadn't been in the Army that long, but I knew if I wanted to get anything done, the NCOs were the ones to go to first. Not getting assigned to SF, among other things, meant not going back on jump status (read that: not getting the extra $110 a month for jump pay). Polly would be hard pressed to make it in graduate school with only about $275 a month.

After introducing myself, I sat down and said, "I hear that some of us with orders to the 5th Group are being syphoned off to other units."

The grisly old master sergeant nodded and mentioned something about "Needs of the Army."

"Yeah. That may be so, but I was sent to the Vietnamese Language School en route to the 5th Group." I unfolded a copy of my orders and gave it to him. "It'd sure be a shame not to be able to use my language ability. Not to mention a real waste of the Army's money."

He looked at the orders and slowly nodded his head. "You may have something there."

He didn't ask if I had gotten a rating in language school, but I was ready to lie about my language proficiency and throw out some handy phrases I had remembered like "I'll have another cold beer please." and "Where is the toilet?" and "You are a pretty girl!" Fortunately, it never got that far, and three days later I was on a plane heading for the 5th Group headquarters in Nha Trang.

5th Group headquarters in Nha Trang

I SAT AROUND the headquarters for about a week waiting to get my assignment within the Group, enjoying the mess hall and the All Ranks Club. Speaking of the mess hall: on every table, along with the salt, pepper, sugar, and Tabasco sauce, were bottles filled with the antimalarial pills. One bottle held the little white one that we had to take daily, and another bottle held the big orange ones that we took weekly. I had started them two weeks before I left the States, and the full effect was beginning to take hold. One of the pill's side effects was that it really loosened your bowels. In fact, I don't believe I had a dry fart that whole year.

Finally, I found out that I would be going to an A Team down in III Corps. ODA 325, at Duc Hue. It sat along the Cambodian border tucked in the armpit of the angel wing (see map at the beginning of this chapter–arrow 1). This being the case, I was scheduled for refresher training on the weapons that were found on a typical A site/camp: 30 cal and M60 machine guns, 60 and 81 MM mortars, M16 and Car15 automatic rifles, M79 grenade launchers and so on.

The training would take place on Hon Tre Island, about a mile off the coast of Nha Trang. To get there, we boarded a Navy LCM (Landing Craft Mechanized). Its 2 Detroit 12V-71 diesel engines pushed us along at a slow speed, but did the trick. We spent two weeks being brought up to speed in the weapon systems and the procedures for calling in tactical air strikes, helicopter gunships, and MEDEVAC choppers. All in all, it was excellent training and very relevant to what was to come. At the end of training, we returned to the Group headquarters.

I WAS SHIPPED south through our B-Team, finally arriving at ODA 325 Duc Hue. At the time, the 5th Group was set up with two levels of command before reaching the A Detachment. Our C-Team, C-3, headquarters in Bien Hoa; under it was the B Team, B-32, headquartered in Tay Ninh. Later in my career the C-Team would be designated as a battalion, and the B-Team would be designated as a Company. This was done to better align Special Forces with the conventional army. This alignment assisted in the career progression of officers and NCOs as command of battalions and groups fell under a centralized selection process.

ODA 325 Duc Hue

(see map at the beginning of this chapter–arrow 1)

AUTHOR's NOTE: The Special Forces 12-man Operational Detachment A sometimes went by A-Team, ODA, or just plain Team. It's the basic unit in SF, the workhorse if you will, where real men did real things. In Vietnam, the place where the ODAs worked and lived was referred to simply as "The Camp," "A Camp" or "A Site."

THE ONLY WAY in or out of the camp was by air: either helicopter or the fixed wing, twin engine DHC-4 Caribou aircraft. Just outside the camp lay a helipad and a 1000-foot runway, both paved with PSP (Perforated Steel Planking). The camp sat in the middle of a free-fire zone. A free-fire zone is an area where anyone passing through not cleared by the ODA was considered the enemy and could be engaged at will. Old deserted rice patties dotted with tree squares ruled the horizon. A tree square is just what it sounds like: a group of trees growing in a square on built-up ground. In the middle of the square, the rice farmer built a little house for himself and his family. These houses had long since passed into history. These dry oases were the only places where we (or the bad guys) could hold up during the day or night and expect to be somewhat dry. Many a fire fight occurred around these tree squares. When we moved into one, we could often see old Napalm hanging like blackened candle wax in the trees. A monument to more dangerous times.

We conducted interdiction operations along the Ho Chí Minh Trail, a logistical system running from North Vietnam to South Vietnam through the neighboring countries of Laos and Cambodia. The system provided support, in the form of manpower and material, to the Viet Cong (VC) as well as to the regular North Vietnamese Army units operating in the northern part of South Vietnam. It served as a critical strategic supply route for the North Vietnamese forces.

The Civilian Irregular Defense Force (CIDG) with Vietnamese Special Forces teams or Luc Luong Dac Biet (LLDB) and US Army

Special Forces A Teams manned Camps like ODA 325 at Duc Hue. The CIDG were basically mercenaries who lived with their families in the camp. The US Special Forces trained, equipped, and fed them and their families. Having their families there proved to be a good thing in that it basically ensured that when the VC or North Vietnamese Army (NVA) attacked a camp, the CIDG would stay and fight.

THE UH-1H CHOPPER I rode in settled down on the helipad, blowing fine sand and bending grass in all directions. The Team Commander, CPT Robert Romero, stood holding his beret on his head, leaning into the rotowash. Romero, a half-blooded American Indian from somewhere out West, wanted to return to the States and marry the Chief's daughter. And I later heard he did.

Romero welcomed me, and I walked with him through the gate and into the camp, which was only a few meters from where I landed. Here I met the rest of the motley crew of ODA 325. All were wearing Tiger Fatigues, the uniform that SF issued to the CIDG for field operations.

At the time, the ODA had three slots for officers: Commander, XO, and CA/PO (Civil Affairs/Physical Operations Officer). I had been slotted as the CA/PO, pending departure of the XO. It didn't matter what our titles were: we all did the same job when it came to field work.

Life in the camp was similar to life in a small village. We had a mess hall; a medical clinic manned by our senior medic, SSG Jungling, a sharp-faced blond guy with a deep tan from days spent under a tropical sun; a parade field (not that we ever used it); and "homes" in the form of sandbagged bunkers topped with corrugated tin. These bunkers helped define the defensive walls of our star-shaped camp. Fifty-five gallon drums lay aimlessly around, acting as cook stoves, water containers, and support for squares of wood on which rice dried. Coils of concertina wire held in place by engineer stakes and laced with claymore mines lay row upon row to form the camp's outer perimeter.

The CIDG and their families cooked over open fires and ate on tables set up outside their sandbagged homes. The US Special Forces paid the CIDG and also brought in baskets of fish, vegetables, squash, and gallon plastic jugs filled with Nuoc Man, a spicy dipping sauce (extract) made from raw fish. Of course, we enjoyed local delicacies like seven-foot cobras, which were caught, killed, then eaten.

Doc Jungling, the senior medic on the Team, was the most knowledgeable in his field and not only provided medical support to the

Team, but he also supervised Vietnamese medical personnel he had trained to take care of the CIDG and their families.

The young wife of one of the CIDG was going to have a baby. It wasn't supposed to be a big deal as Jungling and his trained Vietnamese medics delivered babies all the time. As luck would have it, the birth wasn't going as well as it should have, and we called for a MEDEVAC.

Several of us helped Jungling get the girl to the helipad. It would rain, then clear up as a thunder head passed over. As she screamed and Jungling worked on her, I held the corner of a poncho that seemed to weigh fifty pounds. I closed my eyes as the blood drained from my head, and my knees quivered. I prayed for her, and I prayed that I wouldn't pass out. I would've never lived that one down.

It was a breached birth, on a helipad, in the rain, under a poncho, and Jungling pulled it off, or maybe I should say "pulled it out," just as the chopper touched down. He, the new mother, and her husband, who was holding a corner of the poncho and trying, like me, to stay conscious, climbed into the chopper and it lifted off. Mother and baby would be just fine. Then and there I realized that there was just about nothing that an SF soldier couldn't and wouldn't do.

Our Teamhouse was constructed mostly underground, with walls of sandbags protecting it from incoming indirect fire. I moved into a room recently vacated by another soldier who had DEROSed (Date Estimated to Return from Overseas). It was fully furnished with an iron cot topped with a thin four-inch mattress, a metal chair, a desk made of old crates that had housed 105MM howitzer rounds, a fan, a small lamp, and, last but not least, a wall decorated with rare art from old issues of *Playboy*.

Since most of the Teamhouse was underground or behind layers of sandbags, it was humid and filled with bugs. On my first night, I lay awake in my boxer shorts enduring the oppressive heat and humidity, trying to get to sleep when something ran across my back. I jumped out of bed and grabbed my flashlight just in time to see a three-inch cockroach scuttle into the shadows. Eventually, I got used to sharing my bed with these and other creatures. It worked fine as long as they didn't stop when they crossed my back or legs.

I enjoyed the use of our three-hole outhouse and even mastered the conservative use of toilet paper which was always in short supply.

A few months before I arrived at the camp, President Nixon had authorized a major push into Cambodia. Units all along the border captured scores of weapon cashes and killed many VC. Prior to that very successful invasion, the VC shelled and probed Duc Hue regularly.

We had our own indirect fire weapons: a 4.2 inch mortar with the range of about 4Ks (4000 meters) and a towed 105mm howitzer that could reach out and touch folks up to 11Ks away. These weapons, along with 30 cal and M60 machine guns at each of the five points of our star-shaped camp, provided us with adequate protection. However, since the big push across the border, the camp had not received a single shell or probe. I wrote Dad that whatever else Nixon did, he did a good thing for me when he authorized the invasion of Cambodia.

In addition to ground forces pushing into Cambodia, B52 Stratofortress conducted Arc Light operations, bombing the hell out of the area. The bombings left huge craters which pocked the countryside and made great swimming holes (see cover photo). At night when the Air Force conducted an Arc Light operation, we could hear the rumble in the distance that sounded like thunder (ergo the name Operation Rolling Thunder) except it went on for several minutes. God help anyone anywhere near one of those bombing raids.

COMBAT OPERATIONS CONSISTED of patrols with two US and around eighty or so CIDG. We conducted interdiction operations, seeking out and destroying any enemy personnel found within our free-fire zone. Often I thought that we were doing more seeking and avoiding than anything else.

On my first operation, Romero handed me several cans of mackerel packed in tomato sauce, telling me how great they were. I had packed my ruck sack with Principal Indigenous Rations (PIRs), the green tinfoil bags of dried food that we provided the CIDG for field operations. For that operation I had packed fish, squid, and pork packages. When we stopped at noon for "pok time" (a rest period during the day which lasted sometimes up to three hours), I opened a can of mackerel and took a bite. God, it tasted like slimy raw fish. About the only thing I could eat was rice from the PIRs. It was a hungry five days for me.

As I said, the camp sat in the middle of rice patties. Whenever I stepped off the compound to conduct combat operations, I waded through water thick with long-bladed, brownish-green grass that reached from knees to chest. I've walked through some very tough terrain in my day, but this was the worst. Between the water and grass tangling my legs and the oppressive heat and humidity, I was ready to stop at a tree square, hang my hammock, and rest. The only problem with that plan was the ants that usually covered the tree square since it was the only dry ground around. No amount of bug juice would keep them from crawling from the trees into the hammock.

We would RON (Rest Over Night) in the tree squares; I would set up a poncho, blow up my air mattress, and hang my body mosquito net under the poncho and over the air mattress. Then I would dig a body trench about a foot deep. Of course, the trench filled with water, but should we receive incoming, I could roll over into it, finding some protection. As daylight gave way to twilight, the ants left the scene to be replaced by mosquitoes. I have never seen anything like it. I could lie on my air mattress and watch the outside of my mosquito net turn darker as the infuriating little insects blocked out the twilight. I would tap the net with my finger and a hole of light would appear, then close up again as the mosquitoes returned.

As I have written, contact with the enemy was, at this point in time, hard to come by. In order to be awarded a Combat Infantryman's Badge (CIB) in the 5th Group, you had to receive fire by a 51 caliber round or smaller. Being shelled or setting off a booby trap didn't cut it. It looked like I might finish out a one year tour in Vietnam and never see any action. Fortunately or unfortunately depending on your perspective, this didn't happen.

I was accompanied by SSG Jungling, the Team's senior medic, on the operation which would earn me a CIB. Jungling and I, along with about sixty CIDG, approached (what else) but a tree square to take pok time when all hell broke loose. We fanned out on line and began returning fire. I grabbed the radio and called back to the camp to see if we had any tac air on station and to inform them that we had made contact. I don't recall actually seeing any bad guys that day. There probably weren't more than four or five holed up in the tree square, but I did hear rounds flying overhead, or thought I did. The whole affair only lasted maybe five minutes; then the VC cut and ran, keeping the tree square between us and them. Later on in the day, one of the CIDG tripped a homemade booby-trap made from a mackerel can and was wounded, proving that those cans of mackerel could mean the death of you.

As luck would have it, I stumbled onto a copy of the situation report SITREP in the B-Team's Tactical Operations Center (TOC) of that operation. I kept the yellow flimsy copy just in case anyone ever questioned my orders dated 27 October 1970 awarding me the CIB. Below is how the first paragraph of the SITREP read:

```
FM: CO, B-32 TAY NINH
TO CO, CO A BIEN HOA

C O N F I D E N T I A L / CITE NR 271- //

SUBJECT: SITREP NR 271, 280001H - 282400H SEP 70

PARA I. OPERATIONS SUMMARY:
A. CONTACTS AND INCIDENTS:
(1) FM A-325 AT 280920 SEP 70 OPN P-69A MADE CONTACT AT
XT277036 WITH AN EST VC SQUAD ARMED WITH AK-47'S UNIFORM
UNK. FRDLYS WERE MOVING TO SECURE LZ. ENEMY WAS IN
STATIONARY POSITION. FRDLY INITIATED CONTACT AT 500
METERS. VC BROKE CONTACT AT 280925H DUE TO FRDLY FIRE
SUPERIORITY AND WITHDREW TO THE SOUTH WEST. NEG FRDLY
CAS, VC CAS UNK.
(2) FM A-325 AT 281430H SEP 70 AT XT256036 1 CIDG ON OPN
P70C TRIPPED A BOOBY TRAP CAUSING FRAG WOUNDS IN HIS
LEGS AND FACE. BOOBY TRAP WAS HOMEMADE MACKEREL CAN
TYPE. CIDG WAS MOVED BACK TO VIC XT275035 FOR MEDEVAC.
MEDEVAC COMPLETED 281630H SEP 70.
```

THE WORD CAME down that we would be turning over the camp to Vietnamese Rangers and that the CIDG would be converted to regular soldiers. There would be two US advisors from the U.S. Military Assistance Command, Vietnam, (MACV) assigned to the camp. This was really going to suck for them. All US Special Forces would be assigned elsewhere in country. With these orders in effect, we began sling-loading out equipment under the belly of CH-47s and a large Sky Crane (a huge helicopter that looked like a dragonfly).

It wasn't long before I received orders sending me to B36, the old 3rd Mobil Strike Force headquartered in Long Hai. The 5th Group was pulling out of Vietnam, leaving behind forces that would train Cambodian soldiers to fight the Khmer Rouge, followers of the Communist Party of Kampuchea in Cambodia. We would retain our green berets, but with a different flash. Most importantly, we would still draw jump pay. As a member of the 5th Group in Vietnam, I'd been awarded honorary Vietnamese jump wings. I never made a jump with them. Similarly, I would be awarded honorary Cambodian jump wings.

Thus ended my tour with ODA 325. None the worse for wear, I was eager to face my next challenge. Especially, since the next challenge would find me located along one of the most beautiful stretches of beach in Vietnam.

I was scheduled for refresher training on the weapons that were found on a typical A site/camp: 30 cal and M60 machine guns, 60 and 81 MM mortars, M16 and Car15 automatic rifles, M79 grenade launchers and so on.

To get to Hon Tre Island, we boarded a Navy LCM (Landing Craft Mechanized).

Standing in front of the gate to ODA 235 Duc Hue.

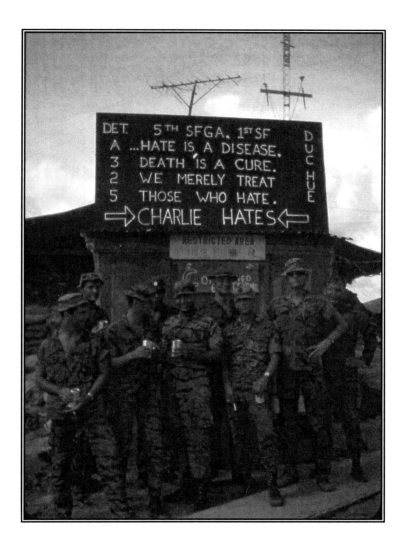

The motley crew of ODA 325.

The only way in or out of the camp was by air: either helicopter or the fixed wing twin engine DHC-4 Caribou aircraft.

A tree square is just what it sounds like. A group of trees growing in a square on built-up ground. This one had a swimming hole, complements of the US Air Force.

Fifty-five gallon drums lay aimlessly around the camp, serving as cooking stoves, water containers, and support for squares of wood on which rice dried.

The Most Fun I Ever Had With My Clothes On 65

The CIDG and their families ate on tables set up outside their sandbagged homes. Concertina wire curled across most of the structures.

The US Special Forces paid the CIDG and brought in baskets of fish, vegetables, squash, and gallon plastic jugs filled with Nuoc Man, a very spicy dipping sauce (extract) made from raw fish.

Doc Jungling holding our mascot, Sally, in front of the Team house.

Of course, there were local delicacies, like seven-foot cobras, which were caught and eaten.

I moved into a room recently vacated by another soldier who had DEROSed (Date Estimated to Return from Overseas).

I enjoyed the use of our three-hole outhouse.

We had our own indirect fire weapons, a 4.2 inch mortar with the range of about 4Ks (4000 meters) and a towed 105mm howitzer that could reach out and touch folks up to 11Ks.

30 cal and M60 machine guns at each of the five points of our star-shaped camp provided us with adequate protection.

The US carpet bombed the hell out of the area. B52 Stratofortress conducted Arc Light operations.

Whenever we stepped off the compound to conduct combat operations, we waded into water and a long bladed grass that reached from knee deep to chest deep.

The Most Fun I Ever Had With My Clothes On   75

I was ready to stop at a tree square, hang up my hammock, and rest.

We began sling loading out equipment under the belly of CH 47s and a large Sky Crane (a huge helicopter that looked like a dragonfly).

B-36 FANK (Forces Armee National Khmer) Training Command

(see map at the beginning of this chapter–arrow 2)

FROM DECEMBER OF 1970 to August of 1971 I spent my time training Cambodian soldiers. We received planeloads of C-130s filled with trainees from twelve to sixty-four years old. I often think about those young boys who had to become men all too soon. The Khmer Rouge, under the leadership of dictator Pol Pot, inflicted a nightmarish slaughter between 1975 and 1979, causing the deaths of at least 1.5 million Cambodians. This blood bath surely cost most of the men and boys we trained their lives.

Up to that point, Special Forces had two missions, Direct Action (DA) and Unconventional Warfare (UW). The mission, to train plane loads of civilians for the Cambodian government, was neither of those. This new mission was to be called Foreign Internal Defense (FID). When the 5th Group left Vietnam and returned to Fort Bragg in March 1971, B-36 at Long Hai was one of a few units that the Group left behind to turn Cambodian civilians into Cambodian soldiers. Also involved in this endeavor were B-43 at Chi Lang and B-51 at Dong Ba Thin. For that next year, these Special Forces training units never exceeded more than six hundred personnel; yet we trained and equipped seventy-eight Cambodian light infantry battalions. Each battalion contained 512 soldiers. I was proud to be a small part of that effort.

I SIGNED INTO B-36 and proceeded to the bunker where I would spend the next eight months. Waiting there for me was a guy who looked no more than eighteen. John Murphy was tall with brown hair and a manner about him that was so mature I dubbed him a young fuddy duddy. Actually, he was the youngest officer in the entire United States Army at the time. He had turned nineteen (the minimum age to hold a commission) just two weeks before graduating OCS. We became good friends and remain so today.

The training we gave the Cambodians mirrored what I had in basic and AIT: small arms, machine gun, indirect fire weapons like the 81, 60, and 4.2mm mortars, hand grenade, river crossing, combat in built up areas, and small unit tactics from squad up to battalion level.

Shortly after I arrived at B-36, I came down with amoebic dysentery, something I must have contracted before I left the A Site. I've had diarrhea in my life, but nothing compared to the gut tearing that I experienced for two days and nights. It was the first and only time that I prayed to die. Finally, I got the medic to give me enough drugs to kill the little pests that were tearing my guts apart.

I HAD PUT in for R&R when I first came in country, with Hawaii being my chosen destination. I'd only been at B-36 about two weeks when I packed up and left.

Polly and I had written back and forth about it for several months, and when I got a free ticket and a reduced rate on hers, that sealed the deal. We were meeting for a five-day stay in Honolulu!

I always hated flying, especially long flights. Polly and I, both tall, always found airplane seats cramped. As I sat in the rear of the aircraft with my knees up around my ears, the stewardess took pity and said that there was an open seat in first class if one of us would like it. I turned to my traveling companion, pulled out a quarter, and flipped it. He won and I lost.

The flight finally landed, and it took me forever to muscle my way from the back of the plane to the door. There waited Polly, all smiles, rested, jumping up and down and waving like she hadn't seen me for six months. Easy for her to do.

We checked into our hotel room sixteen floors up with a view of Waikiki beach. We bought matching blue shirts dotted with little white palms trees, and for the next five days, we lay in the sand, drank colored liquids laced with rum from odd shaped glasses, and, of course, got tickets for the Don Ho show and heard him sing "Tiny Bubbles." On Christmas day, we rented a car and drove around the island stopping for lunch at a little hamburger joint. Money still being a factor.

But all good things must end, and we sat in the airport waiting for her flight back to the states. The call came, and we walked to the boarding check in, tickets in hand. I was antsy to say the least. Polly was trying not to cry. Me, too. I gave her a hug and a goodby kiss and watched her turn and walk toward the plane, glancing back from time to time. This time her seat *was* upgraded. My flight back to Vietnam wouldn't leave for another three hours. *Oh well*, I thought. *Look on the bright side. I'll see her again when my tour is up in six more months.* Or so I hoped.

WHEN I GOT back in country, the monsoons had arrived with a vengeance. I wouldn't see rain like this again until I participated in the Penrod's Triathlon in Fort Lauderdale, Florida. Curtains of rain crossed the compound, drumming the corrugated tin roof that topped our bunker. I lay there listening to the familiar and comfortable sound in an uncomfortable and dangerous land.

I finally settled into the routine of training the Cambodians. My favorite class was the rope bridge. I would show the students how to make a transfer tightening system using climbing ropes and snap-links. To demonstrate the tightening effect, I would pick ten of the largest men and place them on one side of a man-made water obstacle. I would then pick five boys. A tug of war commenced that always ended with the boys pulling the men into the water.

Near our compound stood French chalets, now stripped and ragged skeletons. With their hollow windows and sun-bleached walls, they reminded me of an old scull dug up from some ancient grave. They stood as monuments to a time owned only by history. We used some of these old places to train in urban combat, teaching the students how to breach barriers and enter a room. Was this in violation of the ghosts who lived there? I thought probably so.

WE CONDUCTED LIVE-FIRE exercises where the platoons would maneuver up to and across an objective. During these exercises, the weapons platoon would support with indirect fire, shifting the fire as the maneuver units approached and consolidated on the objective. Once a defective 81mm round landed short. One of my students was hit. I ran over to him. A small hole oozing frothy blood gurgled in the middle of his chest. A sucking chest wound. I wrapped a poncho around him and used my belt to hold it tight. As the whop-whop of the Medevac approached, another instructor and I picked him up. With me holding his back and the other guy holding his feet, we carried him to the waiting chopper.

After it was all over, CPT Marini, my Team leader, walked up and asked if I wanted a medal. Since I had been the one designated to write up all awards for the unit, I said that I didn't need any more work than I had, and no thanks! Later on at the end of my tour, I was told to write myself up for a Bronze Star for service. I was insulted and wouldn't do it. Now I wish I had. Oh well.

AT THE END of training, we turned the students over to the FTX (Field Training Exercise) Committee. This group took the students, now highly trained infantry soldiers, into the jungle where they may or may not have run into bad guys. My roommate, John, was on this committee, and I had to hear about how tough and dangerous it was for them out there. Right. Then someone (a staff officer) had the bright idea that the training cadre, of which I was a member, should, on occasion, and time permitting, deploy on some of these operations.

Since John was currently in the field, I volunteered to spend a few days with the battalion he was "advising." I took off, flying in a UH-1H chopper. Modified M60s hung out each side. A double canopy jungle that stood thick before man walked upright and would stand thick long after he had taken his last step, rushed beneath us. After about an hour, we saw a wisp of purple smoke marking the LZ that the battalion had cleared for us. It was nothing more than a hole in the jungle, just large enough to get a chopper in. Maybe.

We all made fun of chopper pilots and their sacred "crew rest," saying that it wasn't their flight pay we resented; it was their base pay. Anyway, the warrant officer, just a kid, driving this chopper really had his stuff together. He pulled into a hover above the hole and started down. I couldn't believe it. The rotor blade were actually clipping leaves from the trees as we sunk into the jungle. It went from being bright to gradually darker then darker shades of grey the farther down we sank.

Finally, we touched down. I eased off the side and stepped onto the ground, deciding that I would rather walk out than take the chopper back up. The guy lifted off like it was a Sunday drive through the park, then disappeared.

I could see and smell smoke from the campfires just off the LZ hole and knew that John had to be somewhere around. And there he posed, swinging in a hammock, holding a book in one hand and an M16 in the other. A can of Falstaff sat on his stomach. Real combat it was. The boy was well read.

Moving through the jungle was hard, but not as hard as moving through the rice patties. You could be ten feet away from someone and not see him. Every once in a while we would come into an open area that looked as if it had been burned off. Desolate as the moon, the effects of Agent Orange. We walked through these areas and even RONed in or near them. Is it any wonder that I have a prostate the size of a lemon with the consistency of a walnut? Every so often we would stumble onto large mounds of dirt as hard as concrete built by termites. The jungle proved an interesting place.

AT THE END of the final field operation, we had a graduation exercise for the now Cambodian Soldiers. The Battalions stood on the PSP runway, Cambodian flag waving; their gear lay at their feet awaiting a final inspection. Then they loaded on C-130s and flew back to Cambodia, soon to be fighting for their country, but to no avail.

WHEN NOT CONDUCTING training or wandering through the jungle, we had a pretty good life. B-36 at Long Hai sat not a quarter of a mile from one of the most beautiful beaches I had ever seen. And it was only about a forty-five minute jeep drive from Vung Tau, one of the in-country R&R centers. Along the road, large piles of salt harvested from salt water ponds lay capped with woven grass. The occasional pagoda stood off the road. One had a sign in front that read, "**Please do not bomb or shoot on the pagoda**." Made you wonder.

We often ate at a little café, Poinciana, which sat on the beach. It had the best Chinese noodle soup I have ever put in my mouth. We often had our Team parties there. Next to it stood the three story Hotel de la Piscine that rented rooms by the hour or night. The restaurant on the third floor served great seafood. I have never before or since seen shrimp (giant prawn) that big. These suckers looked prehistoric. And of course there was the Vietnamese beer called "*Ba Moui Ba*" which is derived from 3x10+3 = 33. I kept cases of Black Label beer under my cot, bought at the PX in Vung Tau at ten cents a can, but I didn't mind paying twenty-five cents for Bierre 33.

Occasionally, some of us would take a jeep along with our M16s down to the beach and hang out. We would sit, watching glints of sun skip off waves. Every once in a while some of the local girls would show up and pose for photos.

Life in the compound wasn't all that bad either. I can remember sunning on the top of a bunker and drinking champagne. One time I had way too much to drink and fell off. That night, still feeling the effects, I had a night terror: a nightmare on steroids. I vividly remember lying in my bunk with my M16 on the wall to my left when the door opened and in stepped a guy wearing a cone hat and black PJs. His AK-47 at the ready. I remember trying to grab my M16 but being paralyzed. As hard as I tried, I couldn't move a muscle. Then all of a sudden I could! I tore the M16 off the wall, rolled out of the bed, flipped the safety to fully auto, and was about to cut loose when I realized it wasn't real. Lucky for John who was sleeping in the bunk by the door. He never liked being waked up in the middle of the night. After that, when there was only champagne to drink at a party, I volunteered to be designated driver.

Potable water was always at a premium. When we showered, we would wet ourselves down then soap up. Then wash all off. End of shower. Years later when we lived in DeRidder, Louisiana, our septic tank would occasionally back up, and we all would have to shower this way. The kids called it a Vietnam shower. Oh, those were the days.

We also had a marriage counselor. That's right—a marriage counselor. MSG Rufus Redding, but if you knew him well enough, Ready Rufus. His worn body had seen, off and on, five one-year tours in Vietnam. He had suffered through six wives and as many divorces. He was unusually good in that he'd made every mistake you could make in a relationship, and learned from most of them. In our all ranks club, he had a separate little table with two chairs. He held his "counseling" sessions there when the mood struck. Anyone with a problem concerning a wife, girlfriend(s) back home, or girlfriend(s) local could sit with him and unburden themselves. His specialty was the local girls. His only charge was beer. All he could drink during the session. And at 25 cents a can, who couldn't afford it?

SPEAKING OF LOCAL girls, this reminds me of another little story. Many of the guys would meet girls down on the beach. One day CPT Marini and a guy named Fields sat at the Poinciana, the little café and bar we frequented, enjoying the sights and flirting with two of the local girls. After all, the large hotel which rented rooms by the hour or day sat only a block away from our little beachside bar. Some Vietnamese civilians took issue with this, but didn't want to tangle with two guys carrying Ithaca 37, 12-gauge pump-action shotguns loaded with eight rounds of high brass 00 buckshot backed up with Colt Model 1911 45 automatic pistols. Some Vietnamese civilian reported something that brought out the Vietnamese police whom we called White Mice because they wore white helmets.

A jeep carrying four White Mice screeched up to the bar. They climbed out and fanned out into a threading formation, M-16s drawn, heading for Marini and Fields. Instinctively they grabbed their weapon and rolled into prone firing positions behind two palms just as rounds fired by the White Mice whizzed over their heads, or so the story went. Our guys cut loose, killing two and wounding two. The two wounded Mice managed to make it back to their jeep and fishtailed out of the parking area. Fields kept unloading on the jeep as it wheeled around the corner, putting several holes in its back and adding to the injuries the driver and passenger already had.

Later a horde of police descended on the gate to our compound demanding that we turn over Marini and Fields. We wouldn't. In the US 15-6 investigation that followed, Marini and Fields were exonerated, somewhat. The owner of the Poinciana, who received a lot of business from us, backed up Marini and Fields' story. When the investigator asked Fields why he kept firing at the jeep when the two White Mice were driving away, Fields was reported to have said, "Because it was still moving."

Whatcha gonna do?

My tour was drawing to an end. I had requested assignment back to Fort Bragg with either the 5th or the 7th Group. When my orders came down they read Fort Devens, Massachusetts. I was furious. No way was I going north of the Mason-Dixon line. Who did they think they were dealing with here! After the initial shock, I settled down, knowing that I really had no choice. Since I had gone Voluntary Indefinite to get out of the 82nd and into SF, I owed Uncle Sugar one more year. I resolved to give it to him then put in my papers to get out as soon as I could. I'd show 'em!

We received plane loads of C-130s fill with trainees from twelve to sixty-four years old.

The Most Fun I Ever Had With My Clothes On 85

Weapons training included indirect fire weapons like the 81, 60, and 4.2mm mortars as well as the hand grenade.

Polly and I met for R&R in Hawaii.

My favorite class was the rope bridge.

We used some of these old places to train in urban combat, teaching the students how to breach and enter a room.

Near the compound stood old houses built by and for use of the French.

Modified M60s hung out each side of the UH-1H chopper.

The warrant officer, just a kid, driving this chopper really had his shit together.

My roommate, John Murphy, in the field with his battalion of Cambodians.

You could be ten feet away from someone in the jungle and not see them.

Every once in a while we would come into an open area that looked like it had been burned off. Desolate as the moon, the effects of Agent Orange.

Every so often we would stumble onto large mounds of dirt as hard as concrete built by termites.

Along the road, large piles of salt harvested from salt water ponds stood capped with woven grass.

Occasionally, some of us would take a jeep along with our M16s down to the beach and hang out for hours. Every once in a while some of the local girls would show up and pose for photos.

We would eat at a little café which sat on the beach.

Sunning on the top of a bunker and drinking champagne.

The Battalions stood on the PSP runway, Cambodian flag waving.

I often think about those young boys who had to become men all too soon.

Chapter 3

Fort Devens, Massachusetts

WHEN I LEFT Vietnam it was with the intent to put in my final year and exit this man's Army. To say that I was disappointed to be heading to Fort Devens, Massachusetts, would be a gross understatement. As it turned out and as I look back on my thirty-year career traveling all over the world, I can say without any hesitation that our tour in New England was by far the best time of our lives. It was here that I would realize a boyhood dream. And it was here that I decided to make the Army my career.

I DON'T REMEMBER much about the flight back to the states. I left Vietnam behind with mixed feelings of relief and loss. I was coming back whole and no crazier than when I left, so I had dodged the proverbial bullet, no pun intended. I would always remember that year as one of the best of my thirty.

Polly was about to graduate from the University of Georgia with her masters. I was determined to make it to her graduation, but I also had to get signed into the 10th Group at Fort Devens and get on the housing list. There was no way we could afford to live off post for very long. Not in Massachusetts anyway.

I planned the flight to Boston's Logan Airport and then rented a car to drive the forty-five minutes to Devens. When I walked into the Group headquarters sporting a very fine handlebar moustache and long (by 10th Group and the US Army standards) hair, I grew looks of both humor and disdain. The Group assistant adjutant (S1), a captain, immediately pulled me aside and read me the riot act. The 10th Group was a "high-and-tight" organization, and the long hair and particularly the moustache would not cut it. A skeleton crew ran the show as the Group was deployed to Europe on its annual UW exercise, Flintlock.

Well, I'd be around a little less than a year as my plan was to get out as soon as I was eligible. I told the captain that I was just signing in to get my name on the housing list then back out on a three-week leave to pick up my wife in Georgia. When I signed back off leave, I told him,

I would be in compliance with the high standards of the Group and not before. I didn't put it exactly that way, but he got the message. Most of the guys returning from Vietnam at the time fell far short of Army grooming standards.

There would be a surprise waiting for me when I signed back in off leave.

Within three days, I had my name on the housing list, found a little one-bedroom apartment just off the traffic circle that would rent for three months, then boarded a plane heading to Athens, Georgia. I stayed with Polly at her apartment, enjoying the good life, twirling the handlebars of my moustache, and raising my eyebrows at her every chance I got.

Graduation went off without a hitch, and we packed up in the green Chevy Nova and headed north. As I recall the trip was uneventful. We pulled into Ayer, the town just outside Fort Devens, and drove around the traffic circle, finally pulling into the little place I had found. Polly's reaction was so-so about the place until she opened the cabinet under the sink. There was nothing but dirt! She looked up at me then back under the sink.

"How long?" she said.

"How long, what?" I said.

"How long before we can move on post?" She crossed her arms and sucked in a deep breath.

"I don't know. May be a couple of months." I could tell she wasn't impressed with my apartment hunting, but, hey, it had a bed, indoor toilet, refrigerator, and stove. What more did we need? Right? In the end, she accepted it as the seasoned soldier's wife she would become.

I stayed with Polly at her apartment, enjoying the good life, twirling the handlebars of my moustache, and raising my eyebrows at her every chance I got.

Appalachian Trail Walk

THAT FOLLOWING MONDAY I reported into the Group with short hair minus my beloved moustache. The same guy I had met when I signed in told me to report to 2nd Battalion. From there I was sent to B Company and into the office of the acting commander, Captain Sheldon Royal. Royal was a tall man with too pretty of a face. I didn't give a snappy solute. Just walked in and introduced myself.

"We've been waiting on you," Royal said.

"Well, here I am. Just getting moved into a place off post. Hoping to get into base housing soon." I gave him my best smile.

"Hey, Nelson. Get in here," Royal called to the door.

In walked another captain. He looked at me with hooded eyes and nodded.

"He'll be going with you," Royal said to Nelson. "Rick's the Team leader for our SCUBA Team."

I looked at Nelson then back at Royal. "Going where?"

Royal shifted in his chair and gave me a big smile. "When the Group headquarters left on Flintlock, the commander said that since the 7th Group had a Team walking the Lewis and Clark expedition route that we should send a Team to walk the Appalachian Trail from where it starts in Maine back to here."

I thought I hadn't heard him correctly and gave him a "say that again" look.

"Well, we sent Dick Chamberlain and his Team up there. A couple of days ago one of them, we think it was Filthy Fred Faloon, popped off a smoke grenade and the park service contacted the Group headquarters. Not happy. Anyway, I'm relieving Chamberlain. Nelson's going up there to take his place. You're going to be his XO."

This wasn't computing. "Look. I just got back from Nam. Haven't even got my pay set up. Got a wife out there in an apartment off post who doesn't know what's going on, and haven't drawn any TA50." (TA50 is the equipment issued to soldiers that they take to the field like rucksacks, sleeping bags, ponchos and such.)

"Not a problem; Nelson will scrounge you up some equipment. Right, Nelson?"

Nelson gave a half-hearted shrug. He wasn't all too excited about going up to northern Maine and replacing Chamberlain.

I tried one more thing. "Look. I'm really a captain. My orders just haven't caught up with me yet."

Royal wasn't impressed. "Well, until they do, lieutenant," emphasis on the *lieu*, "you'll be the XO."

Nelson was a really good guy. I came to find out that Chamberlain was an outstanding officer who had left the NCO ranks as a Sergeant First Class–SF medic to take a commission.

THERE WASN'T MUCH I could do about it. I went back to the apartment and broke the news to Polly. She'd been down this road before when I reported into the 82nd and immediately left for the field. As always, she took it like a trooper. And two days later Nelson and I sat in a 3/4-ton truck bouncing our way north.

We linked up with the Team near the Trail. They were waiting on us. Not happy. Not a one. Nelson and Chamberlain were friends. This was not going to be easy.

"You know Royal says I'm to relieve you. I'm supposed to send you back on the 3/4-ton." Nelson threw his hands up in a "what can I say?" motion.

Chamberlain–tall, dark-haired, with penetrating brown eyes–nodded that he understood; then he said, "Screw that chickenshit Royal. I'm not going anywhere. You can take over the Team, but I'm not leaving 'em."

Well that was awkward. I took a few steps back, looking at the guys. They glared in our direction.

"They know the score." Chamberlain indicated the Team. "They'll do what you say, and so will I."

A good lesson in professionalism. Chamberlain had more soldier in his little finger than Royal had in his whole body. But he had more soldier in his little finger than most of us, so I'm not necessarily putting Royal down when I write this. I liked knowing a man of his character.

THE WORD WAS that we wouldn't be coming home until we reached Massachusetts. The trail started in Baxter State Park in the northern part of the state. That portion of the trail had small wooden shelters stationed about one day's walk apart. We would pass two or three of them in a day. The walk was beautiful. If I hadn't left Polly by herself in a strange and unknown situation, I would've really enjoyed the hike.

We arranged it so that the 3/4-ton truck would link up with us every day or two. If in the morning it looked like we would link up with the truck that evening, we would throw our rucksacks in the back and go Hollywood for that day's walk/run.

The Trail wove its way through the mountains and little towns along its way. Of course when we were close to a town, we ate at a local restaurant then hit the bars. Good times were being had by all. But Polly.

An old dirt logging roads crossed the Trail, often leading to towns. One evening when the truck was bringing us back to our camp site, a big bull moose somehow got in front of us. This was the biggest animal I had ever seen. The 3/4-ton truck's cab is a good seven-feet high. This animal's rump was dead in line with the windshield. Its rack extended out to the right and left of the truck's sides. Every time it tried to step off the logging road, the lights blinded him so that he had to continue down the road. It was slow going. Finally, someone suggested that we stop and turn the lights off. We did and waited. When we cut the lights back on, the moose was gone. Problem solved.

Often we met other hikers. One time we gave this kid a lift into town. I sat in the back of the 3/4-ton with the rest of the guys while Nelson sat up front with the driver. Filthy Fred was eyeing the guy as a smile broke across his face. "Hey, Rodgers." Rodgers was the Team's junior commo guy. "Wanna swap spit?" Where upon Filthy Fred bought up a big wad and spit it into his palm. Rodgers did the same; then they each placed their palms on each other's mouths and licked.

I never went in for the swapping spit, opting instead for the SF lick in the ear. Anyway, the kid, who was sitting directly across from Filthy Fred and Rodgers, looked like he was about to jump out the of the truck as it tore down the highway. His eyes filled his face. I was sure he was hearing banjo music. The whole back erupted in laughter. The look on the kid's face was something that I have never forgotten. When we pulled to a stop at a stoplight, the kid grabbed his back pack, jumped down, and started walking off shaking his head and mumbling to himself. I'm pretty sure he's telling and retelling this story even today.

THE 3/4-TON TRUCKS would swap off once a week, bringing C Rations (a box of canned food packed in individual cartons), mail, and anything else the wives wanted to send up. At the end of the second week, the truck brought up my promotion orders. Nelson pinned on my captain's bars by the camp fire while Chamberlain and the rest looked on.

We now had three captains on the Trail. Hey, the Team needed two extra captains like Dolly Parton needs two extra inches on her bust.

I got on the phone and called Royal. I told him that my ass was coming home and if he didn't like it. . . Well, he knew what he could do.

I hated to leave the Team and all the fun we were having, but I had been separated from Polly ever since our little stint in Hawaii and well you know. . ..

I next called Polly to ask her to come pick me up. As always, she saddled up for the mission, and the next day we linked up at a little town in Maine, and back to Devens we drove.

The walk was beautiful. If I hadn't left Polly by herself in a strange and unknown situation, I would have really enjoyed the hike.

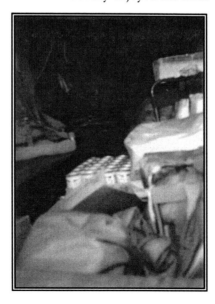

We set it up so that the 3/4-ton truck would link up with us every day or two.

ODA 23/223 Mountain Team

WEEKS WENT BY and the Group and the Battalion I was assigned to returned from Flintlock. LTC McKinney, the 2nd Battalion commander, called me into his office and told me that I would be taking command of ODA 23 in B Company. This was the same company that Royal, who was the company XO, Nelson, and Chamberlain were assigned to.

My company commander, MAJ Painting, welcomed me aboard then told me to report to Master Sergeant Thompson. Thompson was a big burly guy who was extremely intelligent. And I mean that in an academic sense as well as militarily. When I walked into the Team room, he pulled me aside, welcomed me, then said, "Sir, my name is Sergeant Thompson. Just like all the other men you will command on the Team, I worked hard to become a sergeant. I would ask that you respect that and call me and the others by their rank and not our first names."

At the time, it was a habit, and a bad one, for Team leaders to call their men by their first names. I had never done that, but it was prevalent throughout SF. And still is.

Thompson went on saying, "Now this doesn't mean we all can't be friends. Hell, we may become best friends, but we won't and can't become buddies. I need a Team Leader not a Team Mascot. Just so you understand. When the going gets rough and everyone is tired, cold and miserable, the resupply doesn't come and the commo is down and things are basically going to crap all around you and you have to give an order to get up and move out, if you've been trying to be buddy buddy with your soldiers, they just might look you in the eye and say, 'Screw you.' and we don't want that, do we?" He then smiled. I nodded that I understood. And I did.

WITHIN A COUPLE of months after I took command of ODA 23 the Group changed all ODA designations, going from two numbers to three. ODA 23 became ODA 223. I'll explain the logic. The Group had three battalions: 1st Battalion forward deployed to Bad Tolz, Germany, 2nd and 3rd Battalions stationed at Ft Devens with the Group headquarters. The first number in the ODA's designation stood for the battalion. In our case 2nd Battalion. Each battalion had three companies: A, B, and C which corresponded to the second number in the designation, in my case 2 for B company. Each company had five ODA's assigned to it. The third number indicated that we were the third ODA of the five. Well, inquiring minds want to know.

ODA 23 started out as a vanilla ODA, meaning that we had no speciality like HALO (High Altitude Low Opening) or UWO (Underwater Operations or SCUBA). The only speciality training we had had was the same as all other Teams in the group, Winter Warfare. This consisted mostly of cross-country skiing and survival. That would change. My Team sergeant, MSG Thompson, had many talents. One of which was mountaineering. Since the battalion's mission had it inserting its units in the mountains of Europe, someone decided that the battalion needed a Team whose speciality would be mountaineering. Thompson and I lobbied that we should be that Team. Logic prevailed, and under the Thompson's tutelage we began training in earnest.

We trained in class one through five climbs. Our training included not only repelling off cliffs but also from helicopters. We trained not only in free climbing, where we always maintained three points of contact with the wall and climbed without a rope, but also in technical climbing, which involved the use of rope, belays, pitons, and metal spikes (usually steel) that we hammered into a crack or seam in the rock. Our mission was not only to become proficient in the art, but also to teach basic mountaineering to other Teams as needed.

Some of the guys on the Team, like my senior commo NCO, SFC DJ Taylor (eventually retiring as a Command Sergeant Major), who was both HALO and SCUBA qualified with HALO jumps into North Vietnam as part of the 5th Groups CCN program, took to it like a monkey climbing a palm tree. I, on the other hand, lacking upper body strength, looked more like a monkey playing with a football. Well, you get the picture.

THE 10TH GROUP'S AREA of the world at that time was Europe. Every year the Group would deploy on an FTX called Flintlock. It was massive in scale, deploying and controlling units out of a SFOB (Special Forces Operational Base) in England to different locations in Germany, Denmark, Greece, Italy, and so on. Each year the Group rotated between RAF (Royal Air Force) Sculthorp and RAF Greenham Common. This year's Flintlock found us deploying from RAF Greenham Common just west of London. We would go into isolation for a few days then deploy into, normally, Germany, conduct a DA mission, be extracted, have a few days down time, then repeat the cycle.

On our down time, we rented a car and drove all over England, stopping for a picnics along the way. Sometimes we would take the train into London and stay at the British Army's All Ranks club off Hyde

Park. I must have seen *Oh! Calcutta!* four or five times over the period of six years with the 10th Group. We walked through Edinburgh Castle and also saw Loch Ness, but Nessie didn't show.

WE WERE CONSTANTLY undergoing Operational Readiness Testing (ORT). These exercises involved not only direct action (DA) missions but also unconventional warfare exercises, all of which sent us into the mountains of New Hampshire, Vermont, Maine, and across Europe.

When a Team received its mission, they would go into isolation and plan in detail how it would be conducted. Normally, this lasted three to five days depending on the operation. Then at the end they would present a brief back to the battalion commander and his staff. These guys then quizzed the Team members on all aspects of the operation. If all went well, the Team infiltrated into the operational area and executed the mission. I've lost count of the times I went through this process.

On one such Flintlock we deployed from RAF Greenham Common, England, into Germany. My new, XO, LT Tuffs, decided that he would do as he was taught in the SF Course and maintain control of the one-time pads (OTP). One-time pads were crypto pads about the size of a small notebook. They consisted of pages and pages of letters in five groups each. Once they were filled out, they were destroyed. An officer on the Team was supposed to maintain control of them and write and encrypt each message then give it to the commo guy who would send out the encrypted five-letter groups via Morse code over the radio. In reality, the senior commo guy, in this case SFC Taylor, would carry the pads and encrypt the message that I gave him. I didn't feel good about it, but I couldn't tell Tuffs that he couldn't do it as he was taught. Anyway, what could go wrong?

We were three days into the operation. I was sitting by the fire when Tufts walked over, his face colorless. I knew instantly something had really gotten screwed up. "What?" I said expecting the worse.

"Sir, I can't find the one-time pads." He looked down and shook his head.

This was a BIG deal. Not only had we lost a sensitive real world crypto document, we had lost it in a foreign country! "Are you sure?" I said, seeing my career, what little there was of it, flash before my eyes. All he could do was nod his head and gulp shallow breaths.

We went over to his field gear, and I empted his rucksack and turned his sleeping bag inside out. No pads. If we couldn't find them within the next hour, I would send a flash message, in the clear no less, back to our battalion headquarters (called a FOB or Forward Operations Base) to let them know we had really screwed up. There was no question the commander would have administratively yanked us all out of the field and started a 15-6 investigation into the matter.

I stomped back over to my gear and was about to call Thompson over to tell him to get everybody together so we could backtrack where we had come from. Just then Thompson and Taylor walked up. Taylor was smiling. Thompson wasn't.

"Taylor has something to show you." Thompson motioned Taylor forward.

"Look what I found on the ground back at our last stop." Taylor produced the pads.

"You found these back at our RON (Rest Over Night) and are just now telling me?" I knew exactly what he was doing. He was making the point that the commo guy should be the one to control the pads as we, and every other Team, had always done.

Tuffs was right and Taylor was wrong but also right. I called Tuffs over and showed him the pads. His reaction was just like when you've lost your billfold then find it, but multiplied by ten! I allowed as how even though doctrine dictated that an officer control the pads, we'd let SFC Taylor control ours from then on. Tuffs was good with that.

THE NEXT YEAR'S Flintlock we deployed out of RAF Sculthorp. When we went into isolation, the staff presented us with an operations order that tasked us to infiltrate by parachute and deploy ground sensors through the area to monitor railway traffic. The ground sensors would send signals back to our base camp and we would, in turn, encode the time and location and shoot the info back to the SFOB using a coder burst device. All this was well and good, but when we took a look at the sensors, we quickly realized that there was going to be a BIG problem.

There must have been twenty of them ranging in size from a shoe box to an apple crate. Thompson worked out the math and the total load per man ran to around 100 pounds. Obviously we would have to pack them into a bundle and use a cargo chute to drop them. But, no, we were told, the sensors would be delivered to us administratively on the DZ. Something about sensitive equipment that we couldn't afford to damage in a drop? Well, one less thing to worry about.

This whole thing smelled a little off to me. I got the administrative thing with the bundle. But it seemed a lot of trouble to go through not to mention tactically very risky just to get information on rail movements. I would have thought there would have been an easier and safer way to do it. But by now, I'd learned that a mission didn't have to make sense in order to be executed.

We would infiltrate via C-130 and jump onto a DZ near a little German village. Then assemble on the truck carrying the sensors, divide them up between us, and move to a predesignated base site. This was another thing that bothered me. Our field base had been designated in the OPORD (Operations Order) that the SFOB had given us. Normally, we would pick where we set up in an area.

Anyway, personnel from SOCEUR (Special Operations Command Europe) the unit that took operational command of the Group when it deployed, would be setting up the DZ with an inverted L. They would also handle the handoff of the sensors. All this under the watchful eye of a grader.

On the night of 6 June we sat packed in the belly of the C-130 bouncing down the runway soon to be on our way to Germany. As I sat there in the dark listening to the sound of the hydraulics retracting the wheels and the strain of the four engines, I realized that it was D-Day. On that same day from that same airstrip in 1944 Operation Overlord kicked off. And here we were following in their footsteps. Amazing.

THE PLANE LINED up on the DZ, the jumpmaster, MSG Thompson, leaned out the door, his cheeks flapping against the wind. Then came the jumpmaster's commands: "Get Ready" . . . "Stand Up". . . "Hookup". . . "Check Static Lines". . . "Check Equipment". . . "Sound Off For Equipment Check". . . then a ripple of "Okays" came one after another, each followed by a slap on the butt of the guy in front of you. Then came "Stand In The Door." I waddled up with my sixty-pound ruck sack strapped under my reserve, shoved the static line I was holding into Thompson's hand, and placed my hands outside each side of the door. The door light flashed from red to green. The wind whipped around me, its deafening sound almost drowning out Thompson's loud command of "GO." I pushed myself up and out, getting into as tight a body position as I could given the blast of air that lifted and turned me around. Finally, I felt the familiar and comforting tug and heard a pop as my chute opened.

When I looked down, I knew we were in trouble.

I could plainly see that the DZ was off to my left and that the plane dropping the others angled toward the lights of the small German village. This had the makings of a major cluster. I drifted down and off the drop zone, landing on sloping ground off the DZ. I hardly had to do a PLF (parachute landing fall) as I just simply fell forward. While floating down, I had trouble untying the cord that held my rucksack to my leg. Normally, the jumper untied the cord then released his rucksack that was tethered to his parachute harness by a twelve foot lowering line. The cord I had holding the rucksack to my leg had gotten knotted, and I had to ride the rucksack in. On landing, I pulled my knife and cut the cord. In the process I sliced my thumb. It started bleeding like a sieve. Oh, yeah, this was going to be a major cluster.

I fought my way up the slope onto the DZ, stowed my parachute in my kit bag, shouldered my rucksack, hoisted the kit bag on my back, got my bearings, and headed for the rallying point. That was when I heard shouting and lights from the ambulance that always stood by during a drop. Things were becoming admin fast. All semblance of a tactical roll up and assembly forgotten.

When I walked up to the assembly area where we were to turn in our chutes and pick up the sensors, the fingers were already pointing. It seems that half the Team, grader included, landed in the town. The grader had broken his leg in three places and a MEDEVAC was on the way. Three of my guys had badly twisted ankles or wrenched knees and wouldn't be able to continue the mission.

Okay so who screwed up? In a case like that, it's the first question everyone asks. Did the jumpmaster put us out too early? Did he put us out too late? Since we went on the green light that the pilot controlled, it seemed unlikely they could pin it on MSG Thompson. Was the DZ set up wrong?

All that didn't matter. At least not now. With three fewer men than we planned for, carrying all our equipment plus the sensors was a non-starter. I huddled with the SOCEUR guys who came up with this mission. They said no problem. They would just truck all the sensors to our designated base camp. Well what the Hell. The facade of a tactical mission was long since peeled from this operation in a big way. Again the smell that something wasn't right with this operation grew stronger.

We finally arrived at the site that the SOCERU guys selected for us and from which we would deploy the sensors. The wooded area was near a road and surrounded by fields. Not a really good hiding place. Again, something wasn't adding up.

By this time it was getting light and we burrowed in as best we could, hoping that we wouldn't be seen from the road or near by fields.

All that day, aggressor vehicles passed by never giving a second look in our direction. What was up?

We spent the next two nights placing the sensors along railroad tracks through the area, never seeing any aggressors. It was if no one was looking for us.

The data we collected from the sensors was massive. SFC Taylor would spend hours encoding the times and locations then sending the information out via a burst transmission to the SFOB. All the while aggressors passed by our little wooded home-away-from-home, seeming not to care.

Finally, the mission ended, and we set up a night LZ for a chopper extraction. All this time we had no grader. Was this going to be a freebie? When the exfil hour arrived, instead of the whop whop of a chopper, I heard the whining grind of a five-ton tuck. It pulled up and the captain driving it told us to get in. We did and were administratively trucked to an air field then flown back to England.

On our return, hands were shaken and backs were slapped. It seemed that we were the heroes of the day. This didn't compute. I came to find out that the whole point of the fiasco was to see if we could set up sensors and send back the data. Nothing else mattered. The aggressors were told not to bother looking for us. The whole thing was an admin drill to help justify the money that the Army would spend on this, the next generation of ground sensors.

As it turned out the fingers finally pointed to the crew from SOCEUR who set up the DZ. We were free to take off and tour the English countryside until our next insertion. And we did.

WE HADN'T BEEN back two weeks from Flintlock when MSG Thompson, always ready to take on new challenges, suggested that we ask to be trained in the employment of the SADM (Small Atomic Demolitions Munitions) device. Known by the code name "Green Light." The device could be put in a large rucksack and carried on a mission. Sounded good to me, so I approached my company commander, MAJ Painting. He thought it would be a good idea and sent the request up the chain of command. The Group had, at the time, a small training facility manned by the SADM Committee that trained and tested selected members of A Teams on this device. I have to admit that I didn't know what I was getting myself and my Team into. Had I thought it through, I would have never mentioned it, thus saving me and the Team untold amounts of pain in the ass.

The SADM mission required that an A Team have three people trained in the use of the munition, the reason being that at least two members had to be with the device at all times. Like a dumbass, I picked myself, my Team sergeant, and my intel sergeant, SFC Hugh Fairley. A mistake that would later come back to haunt me.

When I write "pain in the ass" I'm not kidding. All members of the SADM Team or Green Light Team had to be in the PRP (Personal Readiness Program), requiring an annual class II flight physical and an upgraded security clearance. This wasn't the problem. The real problem came in the form of unannounced and announced TPIs (Technical Proficiency Inspection). These little babies required Green Light Team members (SADM qualified personnel) to demonstrate that they knew how to employ the device while some technician stood by holding a clipboard evaluating our performance. It turned out that if you didn't pass one of these inspections, the command would relieve you on the spot. If you passed the inspection, that was fine, and you would be given another one within the next six months or so to do it again. The risk vs. reward just didn't balance out. Not in my mind, anyway. But I had jumped out of the wood line and now the Team was stuck with it.

I remember one TPI we went through. The SADM Committee placed us in a room with no windows, and the grader cut the lights out and told us to arm the device. Of course, this caused a major problem as no one had thought to bring a flashlight. When challenged about this by the grader, I told him that if we were going to have to deploy the device at night we would have always have a flashlight. I went on to say that I thought it was a bit half-assed of him to set us up like that. He asked if we would care to improvise, and I told him not even a little bit. He wanted to see if we would use a lighter or strike a match which would have been an automatic failure.

Another thing that bothered us was the fact that no one could see any way that a twelve-man A Team would be granted nuclear release authority. The idea was that the Team could use the device to take out targets like dams, large bridges, electrical grid stations, or other strategic targets where conventional demolition like C4 just wouldn't cut it. But if the powers-that-be wouldn't authorize this, what was the sense in maintaining this capability?

One day I got called into MAJ Painting's office. He told me that my Team had been "selected" to undergo a TPI in conjunction with an ORT. Well, crap! He said that this would be the first time a Team had ever been tested with the SADM under field conditions. Well, crap, again!

We spent four days in isolation, planning the mission and building a mock-up of the target. On the morning of the fifth day, we conducted a brief back to the command and staff along with a rep from the SADM Committee and some "civilian" I'd never seen. And never been introduced to. All asked the usual questions. The "civilian" didn't say a thing. We did get some good questions from the SADM rep. Answered them all and were deemed fit to execute the mission. After the morning briefback, we prepared our personal and Team gear as well as the SADM for that night's infiltration.

So here is how it was to go. We would jump the SADM training device in at night, cash the large canister that protected it when jumping along with our parachutes (read that administratively turn over our parachutes and SADM's protective container to the DZ party), and move cross country several miles to attack a simulated heavy water plant behind enemy lines. Things got complicated from the start because two of the top three men on the Team, myself, the Team sergeant, and the intel sergeant, had to be awake and with the device at all times. This all happened with a grader constantly looking over our shoulders.

I decided that SSF Fairly, the intelligence sergeant, would jump in the device and the three of us would rotate carrying it through the New Hampshire mountains. I think the sucker weighed somewhere between seventy and eighty pounds.

After three nights humping the SADM, we finally made it to the target. I did a leader's recon and left two guys to keep eyes on the target for the fourth night. We planned to hit it the fifth night. As it turned out, I wound up pulling the surveillance guys off the target late that afternoon. Big mistake.

When we infiltrated the target that fifth night. The aggressors guarding it were gone! They had, for some reason, been pulled out late that afternoon. We slipped in and left the device along with the grader where we had planned. All went well. We moved up wind from the target and waited for the simulated explosion. This was required because if the device didn't explode, we were to retrieve it. Right. The given amount of time passed. We heard the artillery simulate go off that the grader ignited. We then moved to our exfil point and were extracted by chopper.

We passed the TPI with flying colors. Later in the week I was called to the Group Commander's office to meet with a panel of officers. The panel was deciding which Team would receive this year's Major Larry A. Thorne Trophy. This was an annual award given to the A Team whose performance demonstrated that it was the best trained in the entire Group. I wasn't aware we were in the running. But what the hell. I wasn't aware of much of what went on above the A Team level anyway.

The group commander, XO, and two battalion commanders sat at a long wooden table. The XO said something to the effect: "We've got to decide which Team in the group will receive this year's award for training excellence. Your Team is at the top of the list, but we have one question. When you guys placed the SADM at the heavy water plant, did you know in advance that the target was unmanned?"

Answering truthfully would wind up any chance of receiving the award, but what choice did I have? I said, "No sir, I didn't. I had pulled my surveillance just a few hours before we went in."

He thanked me for my honesty. We didn't receive the award. But the next year, things would be different.

AUTHOR'S NOTE: In 2013 I received an email from David Brown who was co-authoring an article for *Foreign Policy* magazine. He wanted to interview me. The article which appeared in the January 2014 issue of the magazine documented the Special Forces employment of the SADM Device. He led the article with my Team's exploits in this first ever SADM TPI under field conditions. I was proud to be included as part of his research.

POLLY AND I loved our tour at Devens. We skied the mountains of Maine and New Hampshire, SCUBA dove for lobster off the coast of Gloucester and Rockport, walked Freedom Trail whenever anyone came up to visit, ate in our favorite restaurant over a pub on Harvard Square, attended many off Broadway plays, visited Norman Rockwell's home town and bought several signed prints, shopped for antiques throughout the little New England towns, and generally had a great time.

I remember the day I got the news. I'd just gotten home from work and walked through the entrance hall toward the living room. Polly sat on the sofa, eyes fixed on nothing across the room. She'd been crying. *Someone's died*, I thought as I rushed over and sat down beside her. "What's wrong?" I put my arm across her shoulders. "Tell me."

"I. . . I. . .," she choked back a sob. "I. . . 'm. . . pregnant!" Then the water really turned on.

I was first stunned, then confused. We'd been using "protection." How could this happen? Then reality sunk in, and I was scared witless!

"Hey. It's going to be fine. Even great! No problem. You'll see."

Yeah, right. No pregnancy's a problem for the husband who doesn't have a thing growing bigger and bigger in his stomach for nine months.

Yes, Thomas "Tee" Hoyt Davis IV was an accident but not a mistake. Nine months later, on a cold day in December of 1972, in a hospital in Harvard, Massachusetts, Tee entered the world, and the world changed for us all that day.

Winter Warfare training consisted mostly of cross country skiing and survival.

The Most Fun I Ever Had With My Clothes On

Our training included not only repelling off cliffs but also from helicopters.

We trained in technical climbing, involving the use of rope, belays, and pitons, a metal spikes (usually steel) that are driven into a crack or seam in the rock with a hammer.

On our down time, we rented a car and drove all over England, stopping for a picnic along the way. Sometimes we would take the train into London.

We would jump the SADM training device in at night and move cross country several miles to attack a simulated heavy water plant behind enemy lines.

The Most Fun I Ever Had With My Clothes On   121

The men of ODA 23/223

Polly and I loved our assignment at Fort Devens. We skied the mountains of Maine and New Hampshire.

We walked Freedom Trail. Seen here the steeple where Paul Revere displayed the lanterns, warning the country of the march of the British troops.

The Most Fun I Ever Had With My Clothes On 123

We dove for lobster off the coast of Gloucester and Rockport.
Polly top left in her wetsuit.

Underwater Operations School Key West, Florida

THE DEPARTMENT OF the Army had sent CTP Rick Nelson, the commander of B Company's SCUBA Team ODA 222 orders to PCS (Permanent Change of Station). This opened up the commander's slot on the SCUBA Team. I hated leaving ODA232, but I wanted a shot at a SCUBA Team. They understood.

Now, I don't know how many young boys nine through twelve ever have the chance to grow up and be what they dreamed of, but I did. At least a little bit. When I was young, I watched a TV program starring Lloyd Bridges, *Sea Hunt*. From the first episode to the last I sat mesmerized in front of the TV and dreamed. I would become a Frogman. School proved very hard for me. I had trouble reading and most of all spelling. To this day I still can't spell–thank God for spell check.

Anyway, Dad used to say to me all the time, "Tom, if you don't start studying and apply yourself, you're never going to get into college." Attending college was a total given in my home.

My reply was always the same, "I don't need to go to college. I'm going to be a Frogman." End of story. I was absolutely sure of it. As I grew older, reality set in and becoming a Frogman faded to the back of my mind, but never quite left.

The Group approved my request to attend the Special Forces Underwater Operations School in Key West, Florida. I had a couple of months at the end of 1972 before I started the course. During that time I prepared myself. This was long before the days that the Group would conduct pre-scuba training. In those days you were on your own, which was one of the reasons why the failure rate was so high.

To start with, Polly and I took a civilian SCUBA course taught by one of my SF buddies and great SCUBA instructor, Tom Long. The course ended with a check out dive in Walden Pond, a small, dirty little pond near Fort Devens. I wasn't impressed, but Polly being an English teacher was for some reason thrilled with the fact. And still is. Go figure. I also took a Senior Life Saving course from the American Red Cross. Additionally, I entered their fifty-mile swim and stay fit program. I completed the distance in fifty days. The one thing I did not do was run very much. Big mistake.

RIGHT AFTER CHRISTMAS of 1972 I flew down to Key West and started the six-week course. It seemed like I always picked the worst time of the year to attend these Hooah courses: AIT at Fort Dix, New Jersey, in the winter; Airborne School at Ft Benning, Georgia, in August; and the Q course at Fort Bragg in the winter. I know what you're asking, "How cold can Key West be?" Well, in January of 1973, it snowed in Orlando, Florida. When you are in and out of the water constantly and the outside temp is in the low 60' s, you get damn cold. I remember that during one phase of pool training one National Guard guy got hyperthermia. He turned blue and was shaking uncontrollably. The instructors pulled him out of the pool, threw him in the showers, and when his body temp reached normal, kicked him out of the course. Oh, well.

There were several exercises that were designed to identify individuals who would break under the stress of underwater operations. One of these was the Harassment Swim. The student sat at the bottom of the pool with his weight belt, double tanks, and double hose regulator on. The instructors would then dive down and begin harassing him. They would start with pulling the student's fins off, then his mask off, then proceed to pull the hose supplying air out of his mouth. If that didn't bring the student to the surface, they'd finally turn his air off. The object was to stay under water and counter all these moves as best you could. Often the student would panic and swim to the surface, and out of the course. In the end the student would eventually have to surface. How long the student could hold out determined how well he did.

My turn came and the instructors started the harassment. I was ready for them. They pulled all the tricks out of their bag to bring me to the surface. Nothing worked. Finally, one of them actually unhooked my regulator from the tanks and took it to the surface. I was now down on the bottom of the pool with only my tanks. So I turned the tanks on. Air flowed in a stream of bubbles out the valve. I cupped my hands around my mouth, catching the bubbles and creating a small pocket of air which I breathed. The instructor could see what I was doing, and after about a minute, he swam down, grabbed me by my inflatable life jacket, and pulled me to the surface. He tried to act pissed at me but couldn't pull it off.

Another drill intended to test how bad we wanted to graduate was the crossovers. Here the students would have to swim underwater from one side of the pool to the other wearing their double tanks and fins but without breathing air from the tanks. As soon as the students came up on the other side of the pool, instructors would let them get a gulp or two of air then send them back underwater to the other side. This went on and on and on. Many a student faced with an unknown number of crossovers just plain quit. It was both physically and psychologically challenging.

Our days started with exercises, mostly stretching, pushups, sit-ups, and flutter kicks. The flutter kick required the student to lie on his back with his hands under the small of the back, lift his legs, lock his knees, point his toes, and begin kicking and kicking and kicking. A standard number was 100 at a time. More, if the student screwed something up.

Then came the run. Short and fast then long and slow, with "fast" and "slow" being relative terms. At the time I was a swimmer, but not a runner. In fact, I and an SSG Erickson were by far the slowest in the class. Constantly dropping out of a run was grounds for failure. We always ran in a column of twos. So what we did, as a group, was to put me and Erickson up front. No matter how loud the instructors screamed at us, no one would take the lead. This meant that no one would have to drop out. It infuriated the instructors, but there was little they could do about it. What? They were going to flunk the entire class. Hardly.

When we weren't in the pool or in the ocean we were in the class room. I was amazed at the academics involved in learning to be a Frogman. Maybe Dad was right.

Another thing that students were graded on was the open-ocean swims. These swims ranged from 1000 meters to 2500 meters conducted both day and night. They began in the bay, ending at the large divers sign (twelve-by-fifteen foot sign with a white diagonal stripe from one corner to the other). The swims made on the surface were graded on time. The ones underwater were graded on time and accuracy.

We had the weekends off. We filled them with bar hopping–Sloppy Joe's, The Green Parrot, Captain Tony's Saloon, and many others. Once we were qualified to "go on air" on our own, we could sign out tanks and buddy dive on the weekend. On one of my weekend dives, I happened onto a giant King Conch. I brought it in and cleaned it out. It was the largest of its kind ever found in that area. In fact it was about twice as big as the one in the Key West museum. Polly and I lugged that thing around for years. Finally, it found a permeant home with our daughter Pollyanna in Atlanta, Georgia.

While I was there, Polly and Tee came down for a visit. Tee was only a month old. We toured the island. Wanting to start Tee out on the right foot, we hit all my favorite bars. We watched the famous sunset.

For Polly, the English major, visiting Hemingway's house was a must. Tee seemed more impressed with The Green Parrot.

    I went after the school with a vengeance. My hard work paid off, and I graduated as an Honor Graduate. COL Joe Love, the Group commander of the 7th Group at the time, attended graduation and presented me with the plaque. Of all the many schools I have attended in my life, this is the one I'm most proud of graduating from. I was, at long last, a Frogman. Hooah!!

For Polly, the English major, visiting Hemingway's house was a must.

My hard work paid off and I graduated as an Honor Graduate. COL Love, the Group commander of the 7th Group at the time, attended graduation and presented me with the plaque.

ODA 222 (SCUBA)

BY THE TIME I returned from Key West, ODA 223 had a new Team Leader. I reported in to my company commander's, MAJ Painting's, office. He sat me down and asked if I would mind having Sergeant First Class Ronald Brockelman, currently the Team's senior commo guy (who would eventually retire as a Sergeant Major), as my Team Sergeant. The slot called for a Master Sergeant, one rank higher than Brockelman's. I had heard of Brockelman before. He was both HALO, SCUBA, and Ranger qualified and had, like SFC Taylor, spent most of his time in Vietnam in CCN, conducting recon missions in North Vietnam (or so the story went). I also heard that he didn't have a problem telling folks like it was and that this had gotten him into hot water on more than one occasion. Well, that would be one thing we had in common anyway.

I allowed as how that would be fine with me. This proved to be the second best decision I had made in my life.

I left the office and crossed the street to enter one of the old two-story WWII barracks. Our Team room was on the first floor. There waiting for me sat Brockelman. He stood up, walked forward, extended his hand and said, "Welcome to the SCUBA Team, Sir." Brockelman was originally from Kansas and still had that Midwestern accent. His dark hair was combed to the side. He sported a mustache barely within the regulation. Dark complected, he wore a permanent tan. When he smiled, crow's feet formed at the corners of his eyes. I instantly knew that this relationship, Team Leader/Team Sergeant, was going to work. And it did.

When I look back on my thirty-one years in the Army, twenty of which I spent in Special Forces, and if I were asked to name the one guy who meant the most to me, I would say without hesitation and with a grateful smile, SFC Ron Brockelman.

WHEN I CAME back from Key West, I was not only a swimming fool; I was a running fool as well. Brockelman was the same. Between us we kept the Team in exceptional shape. When Brockelman interviewed a prospective Team member, he had two questions for which the answer had to be "yes." The first: "Do you like to run and swim?" The next: "Do you like country western music?"

Often Brockelman and I would take the Team on an adventure run. We'd start out from the company area wearing our UDT vest (a life jacket that we always wore that could be inflated in an emergency situation by activating a CO2 cartridge), carrying our fins, and hauling a boom box blasting out Charley Pride, George Jones, or other CW artists of the day. We'd run the mile down to Mirror Lake, hit the water, and swim laps (about a mile total) around the lake, passing the boom box from one to the other. When we completed the swim, we'd put on our running shoes and, still carrying our fins and boom box, head out to complete a four or five mile run. God only knows how fast and far Brockelman could have run had he not been a smoker!

OUR MISSIONS WERE the same as all the other ODAs except we had the added capability of infiltrating on or under water. With this capability came the added task of underwater search and recovery, a mission we took seriously. Since we were stationed in New England, we trained to conduct these searches under ice. Several times during the winter, we would cut a hole in the six inches of ice that covered Mirror Lake to conduct training. We only had 3/4-inch wet suits, so we could only spend about forty-five minutes at a time under the ice.

Once Brockelman, LT Croall, my Team XO (a dark-haired college basketball player who stood six inches taller than me), and one other member of the Team, SGT Brad Hendon, spent three days searching Vermont's 400-acre Marshfield Reservoir. The local fire departments filled our tanks and provided a dive boat while the Red Cross kept us supplied with coffee and sandwiches. All we ever found of the three men who disappeared one night when their boat overturned were a jacket and an anchor.

For our efforts in the search, we received a Letter of Appreciation from Mr. Arvids Lazdins, a family member of one of the missing men. The letter was endorsed down through the chain of command to me and read:

```
1. It is with great pride that I forward to you the
attached Letter of Appreciation from Mr. Lazdins.

2. Mr. Lazdins' letter reveals your handling of a highly
delicate task to have been exceptionally professional,
compassionate and above all comforting to the families
of those lost in the boating accident.

3. While discharging your duties as Search Team Com-
mander, you consistently demonstrated an extraordinary
level of skill and judgement in the numerous, critical
situations inherent with underwater search operations.
```

```
Despite the murky water and large search area (3 miles
by 1 ½ miles), you executed your duties in a highly
commendable manner that not only brought credit on
yourself, but has gone far in enhancing relations with
the local civilian communities.

4. My deepest thanks for the sacrifice of your time and
generosity and for a job well done.
```

NORMALLY, WHEN THE company participated in detail support (a cycle that found Teams doing every thing from post police to being aggressors for field training exercises), we got over by being "on call" for search and recovery missions. However, one time the Battalion tagged us to march in a parade. In the spring, every little township in New England had to have a parade and all wanted a Special Forces Team to show up.

We had never been tasked for this, and when the tasking came down, I complained mightily to my company commander. Little good that did. So off we went, rucksacks packed with a blown up air mattress and shouldering rubber M16s. When we arrived at the town, I checked in with the guy in charge of things; then we all took a seat under a tree and waited. And waited. And waited. Finally, I got up to see what the problem was. The parade should have started two hours ago, and we wanted to get it over with and go home.

"Captain Davis, our fife player hasn't showed up yet." The guy in charge nodded to a group of participants dressed as Revolutionary War soldier, one holding a flag and one a drum. "You don't have anyone who plays the fife do you?"

"I can't imagine that I do," I said then turned to the Team sitting under the tree. "Hey, any of you guys play the fife?"

Sure enough my junior medic raised his hand. "I do."

It was a sight to behold, leading the parade along with two Revolutionary War soldiers marched a soldier wearing a Green Beret. It was at that point I came to realize that on a twelve-man ODA there was someone who could do just about anything.

After the parade, members of the local VFW invited us to the lodge where we were treated like kings. Beer and war stories flowed with fury. We had a ball! The next day when I went into work, I told the company commander that the parade was a real pain in the ass, but seeing as how these taskings couldn't be avoided, I guessed that we would, reluctantly, accept any other parade taskings that came down. And we did.

ON ANOTHER OCCASION, the Group tasked us to conduct a ship bottom search under a large barge that had been turned into a rec center for the town of Pawtucket, Rhode Island. When we got there and checked out the ship, Brockelman and I decided that we had better be the first ones down. The water in the harbor was filthy with mud and silt. Of course, we tied ourselves together with a buddy line. We then tied a 120 foot climbing rope to ourselves with a tender on the other end, ready to pull us in should we get into trouble.

We started at the stern, working our way to the bow. We'd been asked to check and see if there were any stress points like large rocks which might be pressing against the ship's bottom. We each carried a large seal beam underwater flashlight. As soon as we submerged, I knew we were in trouble. Within just a few feet the silt blocked out any light. The only way we could tell up from down was by letting a little water into our mask to feel where it was on our face. I couldn't see the light from the powerful flashlight until I touched it to my face mask. Then I could only see a dull yellow glow. I've never been in water that dark before or since.

In order not to become disoriented and swim head first into the mud covering the river's bottom, we had to hold one hand above us, touching the ship's bottom as we swam toward the bow. Finally, we broke the surface. Both Brockelman and I decided at the same time that we couldn't put any of the other Team members under the ship. We had made it all the way without running into any obstructions, so we declared it safe, packed up our gear, and headed back to Devens with a growing respect for those guys who made a living as salvage divers.

If I had thought that my previous ODA had gotten strange missions, I was in for a real shock with what was to come with the SCUBA Team.

ANYTIME WE WERE given a mission anywhere near the water, I always incorporated some type of surface swim or underwater aspect to it. Sometimes we'd jump into a water DZ. Other times we'd conduct a helocast out the back of a CH47 chopper.

Once we were tasked to take out a road bridge over a river on the coast of Maine. The tides affected the river's current, so we planned to place the demolitions charges under the bridge on the incoming tide. This should counteract the river's outflowing current and create an almost neutral flow. SSG Buco, the Team's senior demo guy, and I would enter the water up river, swim down and place the demolitions, which were packed in rubber, waterproof containers and attached with rope and det cord, around one of the pilings. We'd use simulated Time Pencils, acid-decay delay devices, to set off the explosions.

As usual we had a grader for the ORT, but the guy wasn't about to get into the water with us, especially at night. This proved a benefit that would pay off time and again.

As always, things didn't go as planned, and we didn't catch the incoming tide. In fact the tide was starting to go out when Buco and I entered the river. Immediately, things got hairy. With water rushing us toward the bridge, it was all we could do to stay together. I remember seeing the piling approaching and kicking hard to stay even with Buco who had one of the charges and I the other. We slammed into the concrete and somehow managed to get the charges with the rope in between and around the piling. We didn't have the strength to secure the charges, as the water swept us under the bridge and down the river. I remember hooking my buddy line into Buco and both of us kicking as though our lives depended on it at an angle toward the shore where the Team waited to link up with us.

As we broke from the current and reached shallow water, I felt a sharp pain in my knees. The spines of sea urchins that had somehow found their way up from the ocean, which was only about a quarter of a mile from the bridge, had penetrated my wet suit.

The grader had positioned himself on the bridge, but couldn't see any of what went on underneath it. We had placed the charges, although in reality their placement would not have taken out the bridge even if they had gone off. We felt lucky to just have made contact with the bridge and left the charges. Anyway, since no one could say one way or the other, we received credit for a successful mission. See what I mean about missions for which the Group couldn't find an evaluator who would go the distance with us?

THE MOST PHYSICALLY challenging mission came when the Department of the Army asked the Special Forces School (SWC) at Fort Bragg why, with the SEALS available for waterborne operations, did the Special Forces Groups need an underwater operations capability, i.e. SCUBA Teams.

At the time, the SEALS, who were and still are the undisputed experts in all types of waterborne operations, would operate up to five miles inland. Special Forces Teams were deployed deep behind enemy lines. The answer that SWC wanted to give was that a Special Forces

SCUBA Team could infiltrate underwater and move far beyond the five mile SEALS' limit. Well, just how far was "far beyond"? And to answer that question, SWC turned to the 10th Group who, in turn, turned to us.

The mission would be the ball buster of all ball busters if there ever was one. We were to conduct an underwater infiltration of about 1000 meters from a coast guard cutter in Lake Champlain, then cross 135 miles of Vermont's Green Mountains. And do it in six nights. Note that the straight line distance was 135 miles. This equated to a much farther distance when traveling on foot through the mountains.

We went into isolation to plan the mission. The 1:50,000 map stretched from the middle of the Team room around a corner and almost to the side window. I remember LTC McKinney, our battalion commander, stopping by. He looked at the map, shook his head, and said, "Tom, I'm glad it's you and not me. I don't know if I could've ever done something like this." Often staff officers who attended our briefbacks would come up afterwards and say how they wished they were going on the mission. Sure. I appreciated McKinney's honesty.

One strange thing from the start of this mission was the fact that we hadn't been assigned a grader. Normally, we would have an NCO senior to Brockelman or an officer senior to me to serve as the grader. I was simply told that I was on my honor not to use any transportation other than foot to reach the objective. As usual, the Group couldn't find anyone who would or could keep up with us on what would surely be a six-night marathon from hell.

WE ROLLED OFF the Coast Guard cutter in the early evening and swam to the beach landing site. When we got there, we turned our SCUBA gear over to the admin party, then hit the road.

I was good with a map and compass. In fact, I had maxed every compass course I'd ever been on, but Brockelman was better. He would stay on compass throughout the entire quest.

We planned to move only at night, lying deep in the woods during the day. After the second night, it became apparent that we'd never make it to the target in time, so we started using daylight hours more and more.

We were all in great shape for running and swimming with fins. The problem was that we hadn't done that much walking with a rucksack. We weren't receiving a resupply, so we had to pack all our gear, personal as well as Team, on our backs. The rucks weighed close to seventy pounds. Now, I know that doesn't seem like a heavy load, but humping it across the Green Mountains made it so.

My main problem was blisters. As always, the first place that blisters appeared was between my toes. I would pop them with a needle I always carried, but within an hour the blisters would fill back up and hurt like hell. On a later deployment, a member of the Danish Special Forces would show me a trick to address this problem. But that would be later, and this was now. I rubbed Vaseline between my toes, but that only helped a little bit. Eventually, the blisters started to bleed. This caused concern with infection. Others were having similar problems.

By the third night, self-doubt crept into my mind. When Brockelman left the Senior Commo slot to become the Team Sergeant, I had convinced SFC Taylor, from my old Team, to take over the Senior Commo slot on the SCUBA Team. At some point, I approached Taylor and said, "Whatta ya think? You gonna hang in there?"

Taylor shrugged his shoulders and said, "Sir, I'm going to be with you every step of the way but not one step further."

I smiled and said, "Well, I guess you're going all the way then." Taylor cut a glance at me and flashed his half smile.

I had said it, but I didn't really believe it, so as usual I turned to Brockelman. "How about it Top, will we make it?"

Brockelman winked and said, "Absolutely, sir. We're the SCUBA Team!"

Once a day we would have to send a coded message giving the Battalion our grid coordinates indicating how far we had gone. Of course, we offset the location by a couple of thousand meters. Like all Teams, we didn't trust the battalion not to send aggressors after us just to keep us on our toes. And my toes were hurting so bad that staying on them was out of the question.

We had been walking logging trails and secondary roads at night to move through the Green Mountains. There was absolutely no way to move only cross country. The terrain was just too difficult. This was not out of bounds so far as the rules governing our movement went. We made a tactical decision to sacrifice security for speed. So far it had paid off.

On the fifth night, we moved to within five miles of the objective. All were exhausted and limping along on blistered feet. We had walked and run the whole way, and it had taken a toll on both body and mind.

Our objective was a small air strip that we would attack, disabling the control tower. We reached the objective area early the evening of the sixth night. I took one of my guys with me to do a leader's recon and verify the status of guards on the target. My plan was to leave my guy there, go back and brief the rest of the Team, and then hit the target. Our exfil was to be by chopper from an LZ a few kilometers away. Simple was good.

When we moved into a position to observe the target, I was stunned to see no aggressors guarding it. What was going on? On a closer look, I noticed that a panel van sat near the runway. A single person sat on a small folding stool by a blazing fire heating a cup of something. It was the battalion commander, LTC McKinney.

I was tired, hungry (we were only able to carry one meal a day for the trip), and generally in a pissed off mood. I got up, leaving my guy where we were and walked into the firelight. "Sir, what the hell's going on? Where is everybody?"

McKinney turned and toasted me with his coffee cup. "Tom, we didn't believe there was any way you guys could make it this far, so we didn't lay on any aggressors to guard the target or a chopper for your exfil. Actually, the real mission was to define *how far* a SCUBA Team could make it inland after the infiltration, so we picked a point that we knew you couldn't reach. I guess 135 miles wasn't far enough. You guys are something else!"

Well, I went back and rounded up the Team. We all piled into the van to head back to the battalion area. Everyone was plenty happy that this thing was over. In reality, we weren't sure, after all we had been though, that we could have successfully attacked the target. It was just as well that we didn't have to find out.

I RECEIVED A Letter of Commendation from the Group Commander, COL SF Little (Really, he used SF as his first name. Never did know what the S and F stood for). Anyway, paragraphs one and two are below:

```
1. I would like to take this opportunity to commend you
on your outstanding leadership performance during the
period 4 June 1973 to 9 June 1973. During this period
your detachment, A-222, participated in a Long Range
Infiltration training exercise over a 135 mile area
utilizing both SCUBA and land infiltration techniques.
Your positive attitude, esprit de corps, and aggressive-
ness resulted in your detachment obtaining highly
satisfactory results.
```

2. Training missions such as yours represent unique opportunities, both to you and your detachment. Everyone benefits, especially when the mission is accomplished with the high degree of professional excellence you demonstrated. Your achievement is most significant and one of which you should be justly proud.

As FATE WOULD have it, the Group was in the process of selecting a Team to receive that year's Major Larry A. Thorne award. We won it hands down. Of course, we all acted like it was no big deal. But in our hearts we were very proud of what the award stood for, and forever enjoyed the "bragging rights" that came with it!

AGAIN, I RECEIVED a Letter of Commendation signed by COL SF Little. It read:

1. Operational Detachment A-222, Company B, 2nd Battalion, 10th Special Forces Group is to be commended for the outstanding accomplishment of receiving the 10th Special Forces Group "Major Larry A. Thorne Trophy" for excellence in training.

2. This award is presented for your outstanding accomplishments in training during the past year. You have proven during one annual European exercise and one annual operational readiness test that your detachment is the best in the Group. Your actions serve as an example of professional expertise for which the men of the 10th strive.

3. Your selection to receive the "Major Larry A. Thorne Trophy" is a great achievement of which you can be very proud. It signifies the excellence of your detachment training and pays tribute to your hard work and untiring efforts in pursuit of that excellence. The severe criteria by which units are judged to receive this trophy have placed your detachment solidly among Special Forces' FINEST and is in keeping with our proud heritage and traditions.

4. I take great pleasure and pride in having a detachment with your obvious professional competence within my command. I heartily enjoin you to continue your outstanding work.

SHORTLY AFTER BEING presented with the award, the Battalion's underwater operations requirement moved from B Company to C

Company. Along with it moved me, Brockelman, LT Jim Croall, and our Ops and Intel NCO, SFC Mike Nelson.

ODA 232 (SCUBA)

ODA 232'S AREA of the world was Denmark. Hey, somebody had to do it. I couldn't believe our luck! Flintlock was fast approaching, and we would be deploying not to England to be inserted in Germany, but to Augsburg Denmark, the home base of the Danish Special Forces or Jaegerkorps. The Company would set up an AOB (Advanced Operations Base) headed by my new company commander, MAJ Keaney, and deploy its Teams from there throughout Denmark on DA (strike and recon) missions.

That Flintlock we were the first Team to deploy and the last to return. At the end of operations, half of us would attend the Belgium Commando School, and the other half would attend the Danish Combat Swimmer's School taught by the Danish Frogmankorps (Denmark's SEALS).

While under the control of the AOB, we conducted a total of six, three- to five-day, deployments, a mixture of recon and strike operations. For each one we had a Danish Jeager from their Jægerkorpset attached to the Team. Our Jaeger was a sergeant named Preven Jorgensen.

Preven was the guy who clued me in on how to deal with blisters. For the rest of my career, his trick saved me from the annoying problem with blisters in the field or on the Triathlon circuit. I passed it on to numerous people to include my son and daughter.

Here's what he told me to do: take a needle, some cotton thread, fingernail clippers, and a small bottle of alcohol (I used a Tabasco bottle since it was small and hard to break). When a blister forms, dip the needle with about six inches of cotton thread into the alcohol then thread the blister where it forms on the skin and pass it through to the other side and out the skin. The water inside the blister will seep out. Once that is done, clip the thread with the fingernail clippers, leaving a quarter to half an inch of thread on either side of the blister. Leaving the thread in the blister will wick the blister, preventing it from filling back up and becoming sore. About once or twice a day, you can slightly move the thread back and forth. This will break the seal in the event one is formed. Within two or three days, the blister will dry, and you can remove the thread. The most important thing is that you can walk with your blisters threaded this way without pain. Or without that much pain.

WELL, BACK TO my story. I'll relate one of the missions we went on in Denmark that I vividly remember.

The mission was to recon a communications station (actually a telephone company headquarters) on one of the larger islands off the Danish coast. The coastline consists of thousands of islands, large and small, trapped in a web of fjords. We knew the Danes would let the Homeguard (citizen soldiers similar to our National Guard) know that we were in the area. In studying aerial photos, we noticed a small island sitting about 1500 meters off the coast of the larger island where the target sat. It seemed perfect in that it appeared to be covered with trees and had only one house with a small storage shed sitting near it.

We decided to infiltrate via a Danish coast guard cutter as the mother ship and then use a small Zodiac (rubber boat that would hold a coxswain and up to seven men with equipment) to weave our way through the fjords to the small island. The Danes would drop us off and pick us up four days later. The split Team consisted of me, my Team Sergeant Brockelman, four other Team members, and our Danish Special Forces liaison, Preven. On the second night, Preven and I would don Viking dry suits and swim from the small island to the larger one, take photos of the target, and swim back.

The infiltration went well, and eventually the Zodiac nudged up to the small island. Its beach consisted of rocks, and ran only about ten feet, rising sharply to about six feet, then leveling off. I sent two Team members to the left and right down the beach to set up security, and I climbed up the steep embankment and peeked over it. Instead of trees, it was flat as a pancake and nothing but pasture! I turned and went back to the boat to tell the Dane that piloted the Zodiac that he had the wrong island. He said that this was the island we had chosen, and he had to leave. He immediately shoved off. We later discovered that the black and white aerial photos the AOB had given us were taken of the island after the pasture had been burned off. The burned patterns looked exactly like trees.

Of course, we could see the large house and the toolshed with no problem, so we moved in closer to have a look. Finding it unoccupied, we assumed it to be just a vacation home, used only on weekends. We decided to stay in the tool shed which was about twenty feet long and fifteen feet wide and half filled with junk. It was a tight fit, but the seven of us would make it our home for four nights.

We wanted to make sure that all on the large island where the target was located would be asleep, so we planned to start the swim at midnight, giving us enough time to conduct the recon and get back to the small island just before dawn.

The night after we infiltrated (the second night), Preven and I pulled on the dry suits. I wrapped the Team's little Leica camera in a plastic bag and stuffed it into my dry suit, positioning it at chest level. We swam across the fjord to the larger island, making our way through the countryside and into a small village, where the phone company building sat framed by a row of trees. I snapped several photos of all sides of the building, and we made our way back to where we cached the dry suits.

We swam to within 200 meters of our small island home when the dark began turning grey, and just to our left we saw a man in a boat who appeared to be hunting swans. Preven and I grabbed onto a small rock that just broke the surface, covered our heads with seaweed, and tried to blend in as best we could. It took an hour before the man finally floated out of sight. Even in the dry suits we grew dangerously cold.

Finally, we made it back to the toolshed and crashed for several hours sleep. I woke up with a sick stabbing feeling deep in my gut. I grabbed Brockelman and pulled him to the far end of the shed and said, "I can't remember if I took the damn lens cap off the camera."

Brockelman tapped the heal of his hand against my forehead. "Attention to detail, sir. Attention to detail! Now what are we going to do?"

I threw my hands up. "Go back and do it again. Tonight!"

"OK, but you're gonna have to break it to the Dane." Brockelman, smiling, thumbed in the direction of Preven. He was really having fun with this one. Would I ever hear the end of it? Not.

I walked over to where Preven lay and shook him awake. "Preven, we have to go back and get more photos tonight."

Preven sat up rubbing his eyes. "What? No. No. We have photos. We don't need to go back." He kept shaking his head.

I hated to lie to him, but sometimes you just can't tell your counterpart the whole story. "I discussed it with Brockelman, and we both think that more photos are necessary."

Preven shrugged, lay back down, and said, "Americans."

That night once again, I wrapped the camera in the plastic bag and was about to shove it down the neck of the dry suit when I heard

Brockelman clear his throat. I turned and he made a "give me" motion with his hand. I pulled the lens cap off the camera and handed it to him. "What? You think I'll screw it up again?"

He shoved the cap into his pocket and smiled. "Just a little attention to detail, sir."

The third night's recon went smoothly and on the fourth night we exfiltrated by chopper from our island. When we got back to the FOB, I turned in two rolls of 35 mm film. A couple of days later our AST (Area Specialist Team) NCO brought us a pile of photos and asked why so many photos of the same thing on two different roles. I had taken the lens cap off after all! I told him that I was just being thorough and showing maximum attention to detail as any good Special Forces soldier would. Brockelman, hearing the exchange, smiled and shook his head.

Lesson learned: Always take a camera with a single lense reflex with you when go on a recon.

The Danish Combat Swimmer's School

WHEN THAT YEAR'S Flintlock ended, the rest of the Group returned to Devens, but we stayed in Denmark to attend the Danish Combat Swimmer's Course at a little submarine base near the town of Korsor. At the beginning of Flintlock, another Team attended the same course, and we got the skinny on it. We knew about the running and long swims, but the best piece of intel we gathered concerned the Commander's Run. The commandant of the Danish Frogmankorps was a Lieutenant Commander named Volke. This guy was in really great shape. At least once during the course, he'd lead the run. What he did was take the Team out the front gate and run through the woods, circling back through the gate. The run would be long and fast. When he got back through the gate, the runners would believe that was the end, but he would continue, turning around and back out the gate. This would serve to really stress everyone as they were mentally prepared to stop.

When he pulled that trick on us, we were ready for it. We gave a big shout out of "HOOAH" and charged on. When he led us back in through the gate for the second time, we didn't stop, but waved him to do it again. He'd had all the fun he could stand. And we were plenty glad he had as we'd had all the fun we could stand, too.

We conducted all swims in the Viking dry suit. Basically, we wore long johns and entered the suit through the live rubber that constituted the neck. Once completely in the suit, we clamped a metal ring around our necks then pulled the rubber over the ring. Last came a live rubber hood that we pulled over our heads and over the metal ring, sealing off all of the body but the hands and a small oval around the face. The end result was that water could not get in next to our skin. Of course, sweat or any other liquid couldn't get out. You didn't want to drink coffee prior to a long swim.

One of the instructors' favorite exercises was to have six of us enter the water with a large wooden beam one-foot by two-feet and eight-feet long then swim it parallel to the shore for a mile. When we dragged it out of the water, we had to run the beam back to where we started, all the while dressed head to foot in rubber.

We swam on our backs, hands crossed on our chest, and chins slightly tucked. To navigate, we had to pick a reference point where we were coming from and keep ourselves at the correct angle. We also used our dive compasses if there wasn't a visible reference point available. In SCUBA school, we swam on our sides with one arm extended out to our front, the other down by our side, glancing forward in the direction we swam every so often to get our bearings. For long swims of six miles or more, swimming on our backs proved much easier. We also conducted high speed casts from the back of a Danish Cutter.

At the end of one training day, Brockelman and I sat in our room when I heard someone calling, "Davis! Davis, you son-of-a-bitch! Where the hell are you?"

I looked at Brockelman, and he looked at me. We were on the second floor of the barracks. We stepped over to the window and looked down. There, straddling a big green BMW motorcycle dressed head to toe in black leather sat my first cousin Ben Ash. I was stunned. How the hell had he found me. Better yet, how the hell did he get onto a secure submarine base? He had been stationed in Germany when he got out of the Army. He'd bought the bike and was now cruising all over Europe, seeing the sites. I waved him up, and he joined our motley crew for the weekend.

On the weekends, we would catch a bus into Copenhagen. Of course we saw the Little Mermaid and visited Tivoli Gardens. Ben and I sat outside a café, drinking Tuborg beer.

The three week course was drawing to a close. We only had a short FTX, then the 10-kilometer swim left to go. The night of the FTX, Brockelman and LT Croall got into it, nearly ending in a fist fight. I never did figure out what brought it on, but I was more than a little pissed. I called them both in and threatened to send them home early. It was a hollow threat because doing so would have meant the end of their careers, but I think they weren't sure what I'd do. Neither one had ever seen me that mad. I made them shake hands and apologize to one another. And we never spoke of it again. A few years later after Croall was promoted to captain and took over a SCUBA Team in Panama, he asked Brockelman to be his Team Sergeant. Brockelman agreed. As I understand it, they did exceptionally well together.

We had two British SAS sergeants going through the course with us. Everything you've heard about how tough these guys are, multiply it by five. We were preparing to hit the water to start the final 10K swim, and I could tell one of the SAS guys wasn't quite himself. Brockelman told me that our SAS friend was coming down with pneumonia. He was wheezing and looked pale. I walked up to him and suggested that he sit this swim out. It would mean he'd not graduate, and he'd hear nothing of it.

Brockelman and I were the Team's two strongest swimmers, so we normally were swim buddies. But for that swim, I hooked into the SAS guy. We started out fine, but after a couple of miles, he began to fade. We were lagging back well into the middle of the pack.

The Danes would pull up along side us in their rubber boat and ask if something were wrong. We'd wave them off. They knew about the pneumonia, but wouldn't pull the guy out unless he asked. It eventually got to the point that he was of no help. In fact, he was barely able to kick enough to keep himself parallel to the surface. I finned on, determined to pull both of us through and not be the last ones in.

We finally made it to the end and crawled out of the water. All the SAS guy said to me was, "Sir, you know I'd have done the same for you." No "thank you" was offered, and none was expected. I'd always held the SAS in high regard, but never more so than at that moment.

The course ended, and we flew home to Devens. We felt good about what we had done. Another bear wrestled; another bear pinned. Polly met me at the door of our duplex, Tee on her hip. I immediately reached out to him, but instead of holding out his little arms to me and smiling, he shrunk away, tucking his face into Polly's neck. He didn't recognize me. This moment is forever burned into my brain. Just like when I found out that Kennedy had been assassinated or when I heard on the radio that John Wayne had died.

It's A Small SF World

My SCUBA Team got tagged to set up the DZ for our Battalion's jump. As Team leader, I served as the DZSO (Drop Zone Safety Officer). I set up the inverted L using panels at each point of the L and adjusted it based on wind speed and direction so that the aircraft could line up on it and the jumpmaster could determine his release point to tap his jumpers out.

At the time, I didn't know it, but a Sergeant Major on board was making his last jump with combat equipment. He was the first one out on the first pass. All seemed to exit okay except the first jumper's parachute had a bad May West (one or more static lines were over the chute, keeping it from fully opening). The jumper streamed in and hit hard. Turner DZ had many gullies crisscrossing it where he had landed. Getting the "Cracker Box" (Army ambulance) out to him would be a real trick. But trying to bring him back out in the Box might prove fatal. Word from my medic was that he might have a broken back. I called for a chopper to evacuate him while the C-130 carrying the rest of the jumpers circled the DZ.

I kept getting calls from the plane relaying that the battalion commander was on board and wanted to know when the DZ would be cleared. I had no idea when the chopper would get there or how long it would take the riggers to get photos of the chute and all the other crap that had to be accomplished before I could open the DZ for more jumps. Anyway, after about the third call from the battalion commander, I decided to cancel the airborne operation, sending the C-130 still packed with jumpers back to the airfield.

Needless to say, the Monday Morning Quarterbacks were out in force. When the Group Commander, COL SF Little, heard the story, he told all that Davis was the DZSO and what Davis did, he did in his [COL Little's] name. And any further comments about the drop could be directed to him. There were none.

I'm relating this story to you because back in 2005, I had just finished assisting through my self-publishing company, Old Mountain Press, the publication of *Tales from the Teamhouse Vol. II*, a collection of vignettes by members of an SF Listserv of which I was a member. I got an e-mail from Bill Combs, one of the members, telling me I needed to contact a List member, Rudy Cooper. Rudy had been awarded the Combat Infantryman's Badge three times (WW II, Korea, and Vietnam) and had written his life story. Bill thought I might be of assistance in self-publishing his book.

I contacted Rudy, and he sent me his manuscript. I wanted to get a feel for the writing, so I began reading the first chapter. It began in 1973 at Fort Devens. Hey, I was there then, so my interest was immediately piqued. When I got to the part about Rudy's burning in on the jump, I jumped up and said, "Jesus Christ! I was the DZSO for that jump!"

Well, it just goes to show that in SF you had better make sure the war stories you tell are true because as sure as God made them little green apples, if you tell a story that's not true, someone who was there will be around to call you on it.

Who would have thought back in 1973 that the captain serving as DZSO for Rudy's last jump would, 32 years later, help Rudy publish his story? It truly is a small SF World.

Special Forces Surface Swimmer Infiltration Course

MY THREE YEARS at Devens with the 10th Group were drawing to a close. I had orders to Fort Benning, Georgia, to attend the Infantry Officers Advanced Course with a follow-on class date to Ranger School. Before I'd joined ODA 222, all members who were not Ranger qualified volunteered for Ranger School. Brockelman, of course, graduated as the Honor Graduate. Not being Ranger qualified, I'd always felt lacking.

About two weeks before Polly and I were to pack up and leave Devens, MAJ Keaney called me into his office. The company had been tasked by SWC, the Special Forces School at Fort Bragg, through the Group headquarters to conduct a test of the Viking dry suit in cold weather for a new surface swimmer infiltration course they were about to bring on line. The month was February, and the test would be conducted in Groton, Connecticut. It would involve swimming long distances out in the ocean and in the river as well as a parachute jump off the coast. It would also mean we would be on TDY (Temporary Duty) and draw an extra $22.00 a day! MAJ Keaney knew I was about to leave and was asking if I wanted to conduct this one last hurrah. What was I to do?

I drove home and charged in the front door. Polly stood, surveying the room, thinking how she would plan the upcoming move. At the time, she was one month pregnant with our daughter, Pollyanna, but we didn't know it.

"You're not going to believe this," I said, trying my best to look frustrated but not quite pulling it off.

Polly turned and narrowed her eyes.

I continued, "I've just been tasked to take the Team to Groton, Connecticut. Something about testing dry suits for SWC."

Polly's frown grew darker as Tee cried in the background. Looked like he was coming down with another ear infection.

"No. Wait. I've got this covered," I said, then stepped into the kitchen, grabbed the phone, and called my sister, Marrlee, eight years my junior. She had always been there for me when I needed her. She didn't hesitate, saying "yes." She would be happy to fly up from Georgia and help Polly with the move.

Polly stood in the kitchen's doorway, arms crossed under her chest. She wasn't amused. "*You* want *me* to believe that you're being forced to leave here and go play around with *your* Team while *I* pack up this house, clean and clear quarters, then drive to Georgia? Is that it?"

"Well, when you put it that way. . . .." She wasn't taking this so well. You know, guys can tell about things like that.

I held out my hands, palms up and gave her my best "What can I say?" look.

Her mouth bent further into a stoic frown, then she turned and stomped out of the kitchen. Hey, it was mostly true.

I DON'T KNOW how many of you have ever been to Groton, CT, but it's a Navy town, hosting a large submarine base. The Navy would provide support in the way of safety crafts for our swims. We would swim during the day and party at night. I don't know if Groton has the most Country Western bars of any other city its size, or if Brockelman just had a nose for rooting them out. Anyway, life was good and getting better!

I remember one night we were in a club drinking with several of the Navy guys who were supporting us. One of them turned to me and proposed a trade. He would swap me his leather Navy jacket for my Green Beret and one of the rubber M16s we swam with. I don't think he believed I would give up the rubber M16, but explaining how I happened to lose it on a swim would be no problem. I immediately agreed. When he tried to back out of the deal, his buddies stepped in and shamed him into it. The sleeves were a little too short for me, so I gave it to my little brother, John. He kept it for years, then returned it to me to give to one of my grandsons. The first one to go grow into it will get it.

The ambient temperature was in the high teens with a water temperature in the low 30's. We swam in the Atlantic and Groton's Thames River with the a Navy Landing Craft, Mechanized (LCM) following us in support. This was the same type of ship that I'd taken out to Hon Tre Island when I first arrived in Vietnam.

After a few swims, we discovered we had to wear ¼-inch neoprene mittens rather than the five-finger gloves. When the water hit our faces, it immediately froze. So for protection, we coated our faces with gobs of Vaseline. This way the ice freezing on our faces didn't touch our skin. Of course, our eyebrows caked with ice. Although we wore heavy woolen socks, the cold was so numbing that when we completed a couple of hours swimming with fins, we weren't able to stand, much less walk, for several minutes. Swimming under these severe conditions proved a challenge, to say the least.

On our last 6.5-mile swim down the Thames River, it snowed. We swam in pairs, and, as usual, Brockelman and I swam together. When we finally got to the LCM with our legs numbed by the cold, the Navy guys had to carry us to the rear of the ship. The Navy Chief in charge of the ship said, "This really sucks! You guys are just plain crazy!"

And with a big grin stretching across his face, Brockelman shot back, "Chief, you ain't going to believe this, but this is the most fun I ever had with my clothes on." That Brockelman.

We terminated the testing by conducting a parachute jump into the Atlantic. When I climbed aboard the recovery ship, Fort Devens' post commander leaned over and shook my hand.

AFTER TWO WEEKS (just long enough for Polly and Marrlee to pack up our VW station wagon and head south), the Team returned to Devens. I loaded my little suitcase, signed out of the Group, and pointed my little brown Toyota south.

While I was in the Infantry Officers Advanced Course, a letter of appreciation filtered down to me. It was signed by COL Faistenhammer who, at the time, served as the Deputy Commandant for Combat and Training Developments at the SF School on Fort Bragg. He expressed appreciation for the job we did testing the dry suits. The second paragraph of his letter read:

```
2. The intent of the test was to evaluate surface
swimming techniques and equipment under extreme climatic
```

conditions. To this end, the evaluation was conducted in water ranging from 32 degrees Fahrenheit to 45 degrees Fahrenheit. Air temperature during the period varied from 18 degrees Fahrenheit to 40 degrees Fahrenheit. Application of the wind-chill factor netted air temperatures as low as minus 20 degrees Fahrenheit. The goal of the evaluation was accomplished due to the dedication and positive attitude of the test swimmers. Their continued voluntary participation under what were extremely uncomfortable circumstances is commendable. Their willingness to undertake and accomplish any task is indicative of a high degree of professionalism.

And we got TDY pay of $22.00 a day extra for our efforts! And it was tax free! HOOAH!!

BY THIS TIME, I'd decided that I wanted to make the Army a career. Somehow I had survived two RIFs (Reduction in Force). The only thing that saved me was my college degree. Something most of my SF officer contemporaries didn't have at the time.

I planned to stop in Washington, DC, and sit down with an Infantry Branch assignments officer and have him check my records. My hometown friend, Spurgeon Ambrose, who had four years on me in the Army, told me every time he had a chance that I had to get out of SF and get myself qualified as an Infantry officer. That was if I entertained any thought of an Army career.

When I got to my interview, the assignments officer led me to a tiny room just large enough for an old grey Army desk and two chairs. I sat there while he thumbed through my records, reviewing my assignments and my OERs (Officer Evaluation Report). The further he got in the records, the more pronounced his frown became. He would glance up at me from time to time, then look back down at the records, shaking his head.

"What?" I said.

"I don't know." He tapped his black Army pen on the desk.

"You don't know what?" I was getting the idea that something wasn't right.

"With an assignment history like this, I don't see how the hell you didn't get RIFed."

That got my attention. I had been getting great OERs (except for that stinker in the 82nd). I sat up in my chair. "So whatcha think I ought to do?"

The assignments officer shook his head and smiled. "Well, I think you should keep your hair cut short, but don't buy any more fatigues."

Then he punched open his pen and started writing on a pad of paper. "I'm going to put a note in your file saying that after the Advanced Course, we'll send you to an assignment where there'll be no way for you to weasel your way back into SF."

To this point, I had always done it my way. I now found myself reluctantly in agreement with what the assignments officer had said. I shook my head in the affirmative. This explained why the Army had turned me down for the Regular Army Commission I'd applied for. My first disappointing failure. Spurgeon's words came back, haunting the attic of my mind.

THE NEXT DAY, I resumed the trip south to Fort Benning. I had asked for and had been lucky enough to get assigned "substandard" quarters for our nine-month stay. "Substandard" meant that the place where we would live on post wasn't quite as nice as other quarters, but we would get our housing allotment and not have to pay rent or utilities! Hey, it would have a bed, indoor toilet, refrigerator, and stove. What more did we need? Right?

Anyway, I was looking forward to a fun nine months before I started Ranger School. The Infantry Officers Advanced Course had the reputation of being a cushy assignment where the students could take a deep breath, relax, get to know their families, and generally have a great time.

Not so fast, big guy.

Several times during the winter, we would cut a hole in the six inches of ice that covered Mirror Lake to conduct training. I'm to the left and Brockelman is to the right.

The Group tasked us to conduct a ship bottom search under a large barge that had been turned into a rec center for the town of Pawtucket, RI.

Other times we'd conduct a helocast out the back of a CH-47 chopper.

All swims at the Danish Combat Swimmers Course were conducted in the Viking Dry Suit. Basically, we wore long johns and entered the suit through the live rubber that constituted the neck.

Lastly, came a live rubber hood that we pulled over our head and over the metal ring, sealing off all of the body but the hands and a small oval around the face. From left to right: SGT McEntyre, jr. commo, me, and Lt Crowll.

We also conducted high speed casts from the back of a Danish Cutter.

On the weekends, we would catch a bus into Copenhagen. Of course we saw the Little Mermaid and visited Tivoli Gardens.

Ben and I sat outside a café, drinking Tuborg beer. I'm to the right and Ben's to the left.

The last thing we did testing the dry suits was to conduct a parachute jump into the Atlantic. When I climbed aboard the recovery ship, Ft. Deven's post commander was waiting to shake my hand.

The Most Fun I Ever Had With My Clothes On 157

ODA 232 Above: Top-left to right–SFC Mike A. Nelson, intell; SSG Donald Miller, commo; SGT Coleman B. McEntyre, commo; SFC Ronald D. Brockelman, Team SGT, 1LT James E. Croall, XO; SP4 Jeffrey Prevey. Lt Wpns; SP4 Robert McKague, Lt Wpns (S4). Bottom-left to right–SFC Kenneth Kilmer, Hvy Wpns; SP4 Jeffrey John, Med Ast; CPT Thomas H Davis, CO; SGT Bruce Hazelwood, Ast Intel; SP4 Brett Walker, Med Supervisor; SP4 Ronald Sheckler, Eng.

Chapter 4

Fort Benning, Georgia

The Infantry Officers Advanced Course

I SIGNED INTO the Infantry School at Fort Benning late in February of 1973, looking forward to a relaxing nine months. Vietnam was no more than a distant memory. Soon it would end, and all would be coming home. It just so happened that about the same time I signed into the school, a guy named Brigadier General William R. Richardson assumed the position of Assistant Commandant. He had the crazy idea that since the war in Vietnam would soon be over, that the school should start really teaching the students something about the Army and how is was *supposed* to be run.

What had been billed as a nine-month vacation turned into nine months of studying your ass off. Why did this always happen to me? We all cried like rats eating onions, but the hard work and studying would one day pay off when I attended the Command and General Staff College at Fort Leavenworth, Kansas.

We learned the ins and outs of personnel administration, supply & logistics, intelligence, and tactics (operations & plans). I'd had some great instructors during my SF training, so my expectations were high. MAJs Siegfried and Redmon, instructors in the tactics departments, proved two of the best I'd ever been privileged to study under. Both would become general officers. We learned how to write Operations Orders from Company up to and including Division.

I STUDIED HARD and did quite well. One day we heard that the third and, hopefully, the final RIF was coming down. Every day the students checked their mail box for the dreaded "slip" that would inform them that the Army no longer needed their services. Thank you very much. When the "slips" finally arrived, somehow I'd managed to squeak by again. Others were not so fortunate. The RIF hit the SF and Aviator officers the hardest. I had seen some exceptional warriors go in the past, and this time was no exception. Again, not having a college degree seemed to be the primary factor.

By far the most significant thing that happened to Polly and me at the Advanced Course was the birth of Pollyanna. She came into this world kicking ass and has done so to this day.

NEAR THE END of the course, I received my orders for Germany, exact unit undecided. I'd get a call later on this. Well, the assignments officer said he would make sure I went to somewhere that I couldn't get back into SF. This was supposed to give me the chance to become a "real" Infantry officer. I sat resolved to that fact and accepted my fate with stoic determination.

The call from the unit finally came. I stood in the kitchen and picked up the phone, "Captain Davis." I held the phone to my ear, having no idea who it might be.

"Captain Davis, my name is Captain Sonny Shaffner. I'm the personnel officer for the First Battalion of the 10th Group over here in Bad Tolz, Germany. You're being assigned to us, and I wanted to welcome you to the unit."

I was stunned and excited. I remember wanting to shout, "Thank you, Lord Jesus!!!" but I kept my cool. "Captain Shaffner, you can't imagine how happy I am to hear from you. I'll be coming your way as soon as I finish Ranger School." We talked a little about the unit. He wouldn't commit to what I'd be doing when I got there, and that was okay by me. I was going back to SF! Screw you, Infantry!!

THE NINE MONTHS at the Advanced Course ended, and I graduated somewhere in the top 1/3 of the class. Nothing to brag about. It was over, and I was glad of it. Throughout the course, I kept up my running and incorporated long cross-country rucksack marches, using a map and compass. This was in preparation for Ranger School which was next up on my agenda.

Ranger School

OF COURSE I started Ranger School in November of 1975. As usual, I'd picked the wrong time of the year to attend another HOOAH school. Ranger School was broken down into three phases at the time: Phase I, at Fort Benning, Georgia; Phase II in the mountains of Dahlonega, Georgia; and Phase III in the swamps of Eglin AFB, Florida.

What can I say about Ranger School? Well, it's by far the Army's finest small-unit tactics course. It prepares junior officers and NCOs to be patrol leaders, testing their physical and psychological toughness to the max.

I was in Class 5-75. It was large, consisting mostly of 2nd Lieutenants, the majority of whom were new West Point graduates. These guys were really pretty good considering most had been in the Army only a few months. At thirty, I was the second oldest in the class. I have to admit that even though I'd done most of the stuff we'd be taught, I still worried. In SF we mostly conducted small-unit tactics, but the difference was that we usually got at least five or six hours of sleep in a twenty-four hour period. Not so in Ranger School. We'd be very lucky if we caught a total of three hours within a twenty-four hour period. I really wasn't sure how that would affect me. Also, in SF we usually ate at least twice a day. In Ranger School, I'd be living off one C ration a day while in the field.

On our very first run in Ranger School, I severely twisted my ankle and had to drop out. This was not the way to start the course. The Ranger Instructors (RI) saw it as my being weak. I went to mandatory sick call. The doctor offered me a medical drop. Apparently, this was the honorable way for those who didn't really want to be there to get out. I refused. He gave me a ankle wrap and a butt load of 800MG Motrin. I'd pop the Motrin before and after the runs. It helped, but after a mile, I'd begin to lag behind. This falling back on the runs was something I'd not done for a long, long time and would cost me points at the end of the course.

During each phase we would conduct patrols, rotating patrol leader (PL) positions. A key position within the patrol was the Compass Man. This was the guy who could make or break the PL. The RI graded each PL on every aspect from his issuing of the Patrol Order to his movement to the objective area and his actions on the objective. If the patrol got lost en route to the objective, well. . ..

I'd maxed every compass course I'd ever been on. Ranger School was no exception. The only guy whom I'd served with that was better than me with a map and compass was Brockelman, my Team Sergeant on our two SCUBA Teams. Because I never got lost, every guy who the RI appointed to be the PL further appointed me as his compass man. This was fine with me, but what would happen when I became the PL and could not be on compass?

Well, the answer to that question was we got disorientated. That's a nice way to say we screwed up and got lost. I'd earned the reputation with the students and thankfully with the RIs as the go-to guy for land navigation, so the RI let me slide when I took over the compass while also being the PL. We finally made it to the objective, and I got credit for a successful (although with comments) patrol.

MY CLASS HAD the Christmas break. This was good, and it was bad. It was good because it gave us a chance to rest. I used the time to develop a detailed Ranger patrol order that I could use during the next two phases. The problem with that was that when I pulled it out to use it to write my patrol order, the RI said that I wasn't allowed to have a "cheat sheet" and that I'd have to write it from memory. It proved not to be a waste of time and effort as one day I'd use my notes to put together a book titled *The Patrol Order* and through my distributor, Byrrd Inc., it sold in Army bases all across the US. This was a "How To" book on writing a Ranger Patrol Order. It sold quite well for several years.

I REMEMBER THE mountain phase in Dahlonega, Georgia, as being cold beyond belief. If it hadn't been for the heavy wool ski socks I'd been issued in the 10th Group, I don't think I could have made it. Even with the rubber overshoes, my feet stayed numb most of the time. I remember walking through the mountains. As we climbed higher and higher, the surrounding terrain went from dark brown to gradually getting lighter and lighter until all around us ice hung on the trees and bushes.

One night while I was lying on the outer parameter of our patrol base, a RI came over to my Ranger Buddy and me and asked who was snoring. I looked up at him and said that it wasn't us. After all, I'd heard nothing prior to him walking over.

"Why did you lie to him?" my Ranger Buddy said after the RI had left.

"Whatta ya mean lie? I didn't hear any body snoring."

"Yeah, well I did, and it was *you*." Even in the dark, I could see him smile. I had fallen asleep and dreamed I was awake. It was the strangest feeling I'd ever had. I was absolutely positive that I had been awake the entire time. I later came to find out that this was a common experience among Ranger students.

WE FINALLY MADE it to phase III, the Florida phase. Here we would encounter the Yellow River, which runs through Alabama and Florida. It empties into Blackwater Bay, an arm of Pensacola Bay. The night of the final FTX, a boat dropped our patrol off on the edge of a swamp that lay along the river. To get to the objective, we had to navigate through this swamp then a few more miles to the objective. The PL usually designated the compass man, but this time the RI did. He picked me. I dialed in my underwater wrist compass and headed out. We waded through the swamp, crossing small streams that meandered through the cypress trees. After a while, we'd see "Ranger high ground." I think we wanted to get out of the swamp so bad that in the distance we kept seeing what we thought was the ground rising. The RIs explained that it was something like a mirage.

Finally, we made it out of the swamp then on toward the objective. I was pretty sure I knew where we were. Then, in the distance, we saw the embers from a fire. We had hit the objective dead on. This was when the RI came up to me and said, "I knew there was some reason the other RIs told me to appoint you as compass man. Good job." As it turned out, the RIs didn't like wandering aimlessly around through the swamps any more than the students did.

GRADUATION DAY FINALLY arrived, and we stood in formation to receive our Ranger Tabs. Mother and Dad along with Polly came for the graduation ceremony. I remember that after graduation we all went to the Officer's Club for dinner. I'd been living off of one C ration a day and was always hungry. Dad couldn't believe how much I ate. I'd also developed a huge craving for peanut butter and jelly sandwiches. I couldn't get enough of them, eating them for desert for several years afterwards.

On 13 February 1975, I graduated #32 out of a final total of 169 graduates, which my Senior Tactical Officer, Captain Campbell, described as outstanding. He put that in a letter he wrote to the unit I was being assigned to in Bad Tolz, Germany, 1st Battalion of the 10th SFGA. In this letter, he further cited my grades:

GRADED AREA	POINTS EARNED	POSSIBLE POINTS
Land Navigation	100	100
PT	35	50
Practical Work Exam	85	100
Leadership Patrol Grades	386	500
Tactical Officer's Grade	79	100
Peer Grades	123	100
Spot Reports	+20	(+ or -)
TOTAL	828	1000

    I was particularly proud of my land nav score but most proud of my peer grades. No, the 123 was not my typo. Don't know how I received 123 points out of a possible 100 points, but. . ..

Chapter 5

Flint Kaserne, Bad Tolz, Germany

ODA 3 (HALO)

I WOULD, UNFORTUNATELY for me, be with Polly for our move from the States to Germany. Our first stop was Charleston, South Carolina, where our plane was delayed a day. The Army put us up in a hotel. Pollyanna was barely two months old and cranky. Tee had come down with a fever, and that night we had to soak him in a tub of cool water. He didn't like it one bit, and let us and all in the surrounding rooms know about it. The only bright spot was that Polly's mom and dad had kept our 135-pound black and white Newfoundland, Lance. They would ship him to us when we got settled.

The next day we left Charleston and nine hours later landed in Frankfurt, Germany. We spent the night in the temporary quarters, catching the train for Bad Tolz the next day. By now Tee was coughing to the point that he would throw up. At first several Germans shared the train's small compartment with us. Then when Tee let loose with his coughing fit, the compartment empted like someone had shouted, "FIRE!"

AT LONG LAST, we arrived at Flint Kaserne, home of 1st Battalion 10th Special Forces Group, in Bad Tolz. Our sponsor had a meal in the oven waiting for us. We's been assigned quarters, a large three-story building we called the "beehive" that housed mostly bachelor officers. Just a stairwell. No elevator. Our two-room apartment lay sandwiched between the first and third floors. Polly looked around the place then walked out into the alcove, looking down the flight of stairs she would have to hall every thing up and down and said, "How long?"

"I don't know. I'll put our names on the list for housing in the Kaserne. It won't be long." Hey, it had a bed, an indoor toilet, a refrigerator, and a stove. What more did we need? Right? The one great thing about this place was that I could put an empty bottle of St. Pauli Girl in the window and the beer man would bring up a case of beer to our apartment, picking up the case of empty bottles. Ah, yes. Life was good and getting better!

Bad Tolz sat 59 kilometers south of Munich tucked nicely below the foothills of the Bavarian Alps. We skied in the winter and visited all the sights in the area: Munich's Hofbrauhaus and the Glockenspiel, where we heard the chime and watched the 32 life-sized figures reenact historical Bavarian events. Finally, a golden bird chirped three times ending the show. We also often traveled to Garmisch, the Army's R&R center, and Oberammergau where we saw all the gingerbread houses painted with fairytale murals and attended an off-season production of the Passion Play. Of course, the play was all in German, and I didn't understand a word. We didn't lack for company as we had arrived in a fairyland and one of the most visited parts of Germany. It, in fact, drew guests like a wet lolly pop drew ants. And all the home folks came to see.

IN 1968 WHEN the 10th Group left Germany and relocated to Fort Devens, Massachusetts, they left behind the 1st Battalion. Shortly after this, the 1st Battalion became an autonomous unit, falling under the European Command. It evolved into having a Lieutenant Colonel as the battalion commander, but above him sat a full Colonel who was Special Forces qualified and served as a "Group Commander" and Community Commander for the Kaserne. When I arrived there, the Group commander was Colonel Wereszynski. This command arrangement would always cause problems between the 1st Battalion of the 10th Group and the Group headquarters back at Devens. Eventually this situation would be rectified, but not while I served there.

I REPORTED IN TO the headquarters and introduced myself to CPT Sonny Shaffner. Shaffner was at one time an admin NCO who had gone to OCS and received a commission. At the time, he was the most knowledgeable individual I'd ever met in the field of Personnel Administration. We talked about what jobs might be opening, and I asked to be considered for command of the SCUBA Team. Shaffner told me that that position had just been filled, so "No can do." In fact all the Teams had captains commanding them.

It just so happened that ODA 3, a HALO (High Altitude Low Opening) Team, was about to go to the field, and they didn't have an Intel NCO or XO at the time. Would I like to go with them? Well, I had nothing else to do, so I said, "Sure." What the hell, every unit I'd ever signed in to sent me immediately to the field. Why would this be different?

My next job would be to explain this to Polly. As usual she took the news with a sigh and a shrug of her shoulders, and off I went.

THE TEAM WAS deploying as part of Alpine Friendship, a large Unconventional Warfare (UW) Exercise that our headquarters conducted annually. It involved Special Forces from countries all over Europe. We would jump in by static line, not HALO. ODA 3 didn't like the idea of a static line jump. It worked out good for me. I wasn't HALO qualified and would have had to administratively meet the Team on the DZ if they had conducted a free fall infiltration.

We would link up with guerrillas (we referred to them as Gs) who came from other conventional units in Germany, training them in interdiction operations, then cutting lines of communication throughout the area. Since the Team was short an ops/intel NCO, I filled that position for isolation and the briefback. I had with me all my notes from the Infantry Officers' Advanced Course as well as my notes from Ranger School. I gave a classic intelligence analysis of the enemy capabilities, probable courses of action, and status. All were impressed. Even ODA 3's NCOs.

We infiltrated that night and humped our rucksacks to a location several miles from the drop zone. En route, the temperature dropped like a lead ball in a pond. Around 2400 we entered a dense patch of woods well away from anything human. I hung my poncho between two trees, using bungee cords at each end to hold it up. With old metal orange tent pegs I always carried, I staked the four corners flush with the ground. Next I blew up my air mattress and unrolled my sleeping bag. I had first watch, and half way through an hour of freezing my ass off, huge snowflakes began falling with a fury. SFC Finney relieved me on watch. I trudged back to my poncho hooch and burrowed in for a night's sleep.

I awoke the next morning, blinked twice, and looked around. It was dark and cold. I peaked at the illuminated dial on my watch. It read 0800. What the hell? I rolled over and my face hit the top of my poncho. What the hell, again. When I wiggled my hand out of my sleeping bag and reached up, I found that the poncho ceiling above me almost touched my body. Looking around, I realized that my poncho home was completely covered with snow. It had snowed all night and a foot or more blocked light from all directions.

It was then and there that I decided it was going to be a long, cold two weeks bumbling around through the German countryside.

THE OPERATION WENT well, and we returned to garrison. Shortly after we got back, the battalion commander, LTC Paul Hutton, called me into his office and offered me command of ODA 3. The current commander, Captain Logan Fitch, had been tagged to move up to the Group's Operations section. I'm sure this move was in the works before I went in the field with the Team. My going in with them on the operation had really been to give me a "tooth check," to see how I'd fit in. I'd been found acceptable by all concerned. The only problem was that I wasn't HALO qualified. I'd made around ten free fall jumps when I was in the 82nd Sports Parachute Club, and impressed no one with my ability to get stable.

I'd always been one of the most qualified in my Team's specialty (SCUBA). Now I would be the most *unqualified* and *inexperienced* member of the Team when it came to its speciality, HALO or Military Free Fall. In fact, the guy with the least number of military free fall jumps had about fifty more HALO jumps than I had static line jumps!

Not a problem, Hutton told me. The Team would soon be running a HALO course for the unit, and I could attend it as Team Leader and student.

And what if I couldn't get stable and pass the course. I thought. *Really frigging great.* To say that I was concerned would be a gross understatement.

My next stop after talking with Hutton was my Company Commander, MAJ Frances X. Brennan. Brennan was short, dark-haired, with a permanent tan. A white crescent scar crossed the tip of his nose, a hockey scar. We chatted a while as he passed on his command philosophy. It was pretty standard. Then he picked up a large coffee mug with the Special Forces Crest on its side and the SF moto *De Oppresso Liber* (To Liberate the Oppressed) arched across the top of the crest. He took a sip, set it down, then ended our talk saying, "Tom, I expect all who work for me to manage competing priorities and limited resources to accomplish any mission assigned."

For years in SF, I'd been doing just that, but I'd never heard it put so succinctly into words. Cool. I stole that phrase from him and shamelessly used it the rest of my career.

Fitch had already left the Team and moved to the Operations shop, so after my talk with Brennan, my next stop was the Team room.

ODA 3's Team room sat above the rigger's shed, and unlike other Team rooms which were generally square and small, ODA 3's room was long and wide. Long enough to house the two rigger tables on which they packed their parachutes. Our lockers, where we kept our personal gear, filled the far end. Several desks and chairs sat sprinkled throughout.

I climbed the stairs and entered the Team room. MSG Donahue sat behind his desk thumbing through jump manifests. Donahue was pencil thin with short, greying hair, a weathered face, and china-blue eyes. He was originally from Maine and still had the accent. At the time, he had well over 2,000 free falls and held a USPA (United States Parachute Association) two-digit D skydiving licenses. I'd gotten to know him and the other Team members during the field training exercise and felt I could be honest with him. By now, he knew I'd be his new Team Leader.

"Sergeant Donahue, I need to talk." He nodded and waved me to a chair by his desk. "I commanded a couple of SCUBA Teams before I came here. I was very good when it came to that speciality. I now find that not only am I not any good with this Teams' speciality, I'm not even qualified." He started to say something. I held up my hand. "Let me finish. I'm not afraid of getting hurt. I'm not afraid that I'll get myself killed. What I'm really afraid of is that I'll hold the Team back. I just can't face being the weakest link." I turned my palms up in a "That's it" gesture.

Donahue sat there a minute just looking at me. Then he said, "Sir, you're going to do just fine. I'll see to it." And he did.

ODA 3 BEGAN coordinating the resources required to run a HALO Course at the German Army Air Field at Altenstadt, Germany. I couldn't be a student and the Officer in Charge (OIC), so Fitch assumed that duty, on paper at least. In actuality, Donahue ran the course. He took me under his wing so to speak, keeping an eye on me as I wrestled with my rucksack while falling at 120 miles per hour from 15,000 feet with my nose and mouth covered by an oxygen mask. He'd critiqued me when we were on the ground. Later, I found out that acting as a personal instructor for anyone was something he just didn't do. Guess he felt sorry for me.

We would exit the aircraft from the side door or off the ramp, pulling our ripcord at around 3,000 feet. About twenty five jumps later, the course ended. Certificate in hand, I found myself HALO qualified and the Team Leader of ODA 3. In between field training exercises, the Group Operations section tasked the Team to run the Jumpmaster course for the rest of the unit. Things went well.

SHORTLY AFTER I GRADUATED from the HALO course, a tasking came down through U.S. Army Europe (USAREUR), for the Group to provide Military Free Fall training for the Danish Jaegerkorps. I would be returning to Augsburg, Denmark. Hey, it was a dirty job but. . ..

When we got to Augsburg and met our students, standing in the group was my old swim buddy, Preven Jorgensen. Preven and I had swum back and forth from one island to the other during a Flintlock exercise a couple of years ago. Small world it was.

We jumped from several different aircraft while conducting HALO training for the Danes. I particularly liked jumping from the old McDonnell Douglas C-47. However, most of our jumps were from an MC-130, a C-130 especially modified for Special Operations. Our highest jump was 25,500 feet, falling two minutes at 120 MPH before we pulled our ripcords. I can remember seeing my shadow racing up at me as I plunged through the cloud cover below. Weird. At the end of the course, the Danes awarded us their parachute badge, and we reciprocated by awarding them ours. We had recently conducted parachute operations with the Germans, and I had been awarded their parachute badge as well. I now had jump wings from four different foreign countries.

Again, due mostly to Donahue, the training went exceptionally well. I received a letter of appreciation sent from COL Jensen, commanding officer of the Jaegercorps through the Military Assistance Advisory Group in Denmark, through Seventh Army Europe, and down to our command.

IT WAS WHILE assigned to Bad Tolz that Polly and I met Vaden and Rose Marie Bessent. They would become dear friends who we would serve with off and on during our entire careers. Vaden commanded an Urban Warfare Team. The Team's mission caused them to wear civilian clothes while conducting operations in built-up areas. When they went to the field, they found themselves meandering around Munich or Berlin conducting link ups with assets or executing dead letter drops and other clandestine activities.

I remember one time when my first cousin (more like my sister) Jana Powers and her husband, Sam, came for a visit. We were sitting in the Hofbrauhaus drinking two-fisted steins of beer and eating schnitzel when Vaden walked in and looked around. Polly started to call him over, but I stopped her just in time. He and his Team were on an FTX in Munich. Hey, no sleeping under a poncho and smearing camo on the face for those guys. I pointed out Vaden to Jana and Sam and explained to them what was happening. We all tried to look the other way as Vaden made contact with an asset. Quite a show, if I must say so myself. At least Jana and Sam were impressed.

ONE DAY WHILE sitting around the Team room planning our next HALO jump, Donahue and I got a call from Fitch in the Operations Shop. He had something to tell us, and he couldn't say what over the phone. We trudged up the stairs in the headquarters building and knocked on Fitch's door. "Come in," said a voice from beyond. We pushed our way in.

"Have a seat." Fitch nodded to two chairs in the corner. "I'm going to tell you something that is classified. You can't repeat this to anyone else. At least not for now."

Donnahue and I nodded our agreement.

"The *Reader's Digest* version: SOCEUR (Special Operations Command Europe) has been tasked by the CIA to conduct some classified training with one of our allies here in Europe. The code name for the operation is SPATULE. You and your guys will jump into France and demonstrate to a special group of French Marines how we would teach Gs in an Unconventional War situation. The training will take place near Bayonne. We have a guy on another Team here in the Group that speaks fluent French. He'll train with you and jump in with you. You'll go in static line."

"Any idea when this will happen?" I said.

"Not sure. You'll have to do a site survey with a rep from SOCEUR and the CIA. That'll be within the week. You can't tell anyone (meaning Polly) where you are going or why." That would be no problem as Polly was used to my going places and her not knowing how long I would be gone or where I was going.

THREE DAYS LATER I caught a plane to Stuttgart, Germany, and linked up with a Lieutenant Colonel from SOCEUR. We flew from there to Paris where we met a cultural attache (read that CIA) from the embassy. Okay, James Bond this guy was not. He stood about five feet six inches, round as a basketball, and peered at us through coke-bottle thick, black horn-rimmed glasses. He wore a long white London Fog trench coat. Anyway, he was a nice guy who really knew his way around Paris. The one night we spent in Paris, he took us to a small restaurant that couldn't have been more than twenty feet wide but stretched at least 100 feet long. We ordered Chateaubriand and escargot. Fantastic!

The next day we drove out to a French Army Airfield. The fog hung like a blanket, blocking out everything that was more than ten meters away. Surely they wouldn't take off in this pea soup, but they did. We flew low, following the telephone and power lines. It was a matter of face. The French couldn't let us see that they were afraid to fly.

I, however, had no compunction letting them know that I thought it was stupid to take off under these conditions and said so. We hadn't been in the air more than ten minutes when sanity prevailed, and the pilot turned around, creeping back to the air field. We landed and then piled into a car and drove seven hours to the training area. Two days later the weather cleared, and the French sent a chopper to pick us up. The area where we would conduct the training was on a French military base. It had rifle ranges and demolitions ranges. It also had a large air strip from which we would catch a ride out when the mission was complete.

The mission was basically administrative, conducted in the field. Since no one was supposed to know we were in France, there wouldn't be any aggressors looking for us. The mission's concept was to demonstrate to the French Marines how and what we taught indigenous forces in a UW role. Why they wanted the training or, better yet, what they would do with the training we didn't know and didn't want to know. My guess was that the answers to those two questions were why the mission was classified.

I RETURNED TO Bad Tolz, and we began pre-mission training. We would instruct the Gs on several aspects of UW operations: weapons training, demolition, clandestine communications, the raid and ambush, infiltration and exfiltration techniques, and so on. We even coordinated with the Air Force for a demonstration of the Fulton Recovery System or Sky Hook. This system was developed by inventor Robert Edison Fulton, Jr., for the CIA and Special Ops units to extract high value individuals, with or without their consent, from deep behind enemy lines.

A Sky Hook operation involves using an overall-type harness and a self-inflating balloon which carries an attached heavy nylon lift line.

An MC-130E snags the line with a V-shaped yoke mounted on the plane's nose and the individual is reeled on board. Red flags on the lift line guide the pilot during daylight recoveries; strobe lights on the lift line are used for night recoveries. Of course, we would use a dummy rather than a live person. The recovery kit, which is dropped on the aircraft's first pass over the drop/pickup zone, consists of two large canisters about twice the size of a beer keg filled with helium, the 300-foot nylon rope, and the inflatable balloon. All in all an impressive little set up, and one that we and the CIA thought would show the Gs just how high speed we were.

We had planned for several resupply drops, not only to train the Gs in how we went about setting up the DZ but also how to bring in weapons, demolitions, and additional rations. This meant that our rucks would only weigh between forty and fifty pounds when we jumped in. This would be a blessing when I found out what the French Marine/Gs had waiting for us on the DZ.

WE LOADED INTO an MC-130 at our air strip in Bad Tolz and flew into France. It was about 2100 hours when we jumped over a large DZ. The jump was uneventful, meaning no one broke anything, and all assembled at the turn-in point within thirty minutes. There waiting for us were four Gs, or French Marines. I noticed that they had a large bundle lying near where we'd turned in our chutes.

I sent our on-loan interpreter, SSG Boswell, over to make contact. It seemed as though the Gs had a present for us. Wrapped in a poncho lay a large sheep carcass. It had been skinned and gutted. It had no head, but everything else was there. It weighted a little over fifty pounds. Boswell talked with the Gs for about two minutes then walked over to me and Donahue.

"Those guys," he thumbed in the direction of the four Gs, "brought chow for the camp and want us to haul it."

I turned to Donahue. "Whatta you think?"

"Sir, I think they're screwing with us. But what are we going to do about it?"

Fortunately, we had jumped in light carrying only a couple of days rations and personal gear like ponchos, sleeping bags, and our M16s. The only Team equipment we had was the radios.

About seven weeks ago I had cracked my ankle while making a civilian jump. The worst part of that experience was having to tell Donahue. He still hadn't forgiven me for it. I had gotten the cast off in four weeks, and the next day started swimming a mile a day in the indoor pool. After a couple of weeks, I alternated between light running and swimming. The ankle was tender but had grown pretty strong.

I told Boswell to tell the Gs how much we appreciated the gift and that the Team Leader had decided he would carry the carcass to the base camp. I empted my ruck and divided my stuff between others on the Team. I had a large mountain ruck at the time. Even so about one third of the sheep hung out the top and flopped back and forth as I trudged along behind the Gs who were leading the way to the base camp. The load wasn't all that heavy, but the awkward flipping and flopping made for difficult walking. That seemed to be the point. These guys wanted to know how we'd react to a change in plans, and who would hump the clumsy load.

After about five hours of tramping through the countryside mostly going in a wide circle just to make the trip last longer, the Gs finally led us into their base camp. They had it set up pretty well with good security even though this was an admin operation.

As soon as we walked into the center of the base camp, a G came over to me and took my load, leaving me with a smelly bloody rucksack that I though I'd get rid of on our return to Bad Tolz. They had the animal on a spit over a fire in less than thirty minutes. It was about 0300 hours, and we were all pretty tired. We set up a small perimeter within the base camp, keeping two of us awake at all times as was our SOP (Standard Operating Procedure).

Early the next day as the fifty pounds of sheep slow cooked over a low fire, I met with the G Chief to establish rapport. I surmised that he was a young captain. The higher up in the officer corps they rose, the more pompous a French officer became, something I'd confirm years later in Zaire, Africa. This guy was friendly and eager to see what we would teach him and his men. Over the next two weeks, I came to really like him.

Our training schedule started with the tactics involved in the raid and ambushes and how we would set up and organize resistance fighters to conduct unconventional warfare. We would conduct the demolitions and the weapons training when we received the resupply early in the second week. I half expected to see our "cultural attache" stroll into the base camp at any time dressed in his London Fog, but that never happened.

We had been eating strips of mutton for the past three days. The thing was getting ripe, but we didn't want to offend the Gs by turning up our nose at their generous offer. I later found out that the Gs had all heard that SF guys would eat just about anything. As we continued to slice hunks of meat off the sheep and eat it, they were convinced that what they had heard was true. The temperature was cool-to-cold at night, so I thought the meat would be okay for a few more days. After the fourth day, most of it was gone, and what was left had begun to really ripen. We decided that it was time to ditch the carcass, and we did. I've got to tell you that I haven't eaten mutton since.

ON AROUND THE fifth day we set up to conduct the Fulton Recovery. The operation would go down during daylight and was strictly admin. All the Gs had heard about the Fulton Recovery and were eager to see it.

We set up the DZ and waited for the MC-130 to make its drop. Expectations were high. I began to worry that something might screw up the demonstration. The plane lined up on our inverted L and kicked the bundle which floated down swinging in an unusually wide back and forth ark under the chute. It landed hard. When we broke down the bundle, we discovered that one of the helium container's release knobs had sheared off when it hit the ground. This could prove to be a major problem as the whole operation depended on us getting the balloon completely filled with helium and into the air.

Donahue walked over to see what was going on. "We're screwed," I whispered to him. "The handle's broken off. I can't release the helium."

Donahue turned the canister over for a better look. "Not a problem, sir. Here." He pulled out a pair of pliers and handed them to me. Why the hell he'd thought to bring a pair of pliers with him to the DZ, I'll never know, but he definitely saved us great embarrassment that day. "Attention to detail" is all I could think. Little did I know that he'd soon save my ass again but in a much, much more delicate and potentially explosive situation.

We filled the balloon and let it float up trailed by the nylon rope attached to the dummy. Just as the rope tightened and the MC-130 with its V-shaped yoke mounted on the nose was in bound, Donahue stuffed something into the dummy's jacket and smiled.

When he walked back to where I was standing, I said, "What the hell was that all about?" I nodded to the dummy who was about to be pulled straight up in into the wild blue yonder and then reeled into the belly of the aircraft.

"I thought I'd give our Air Force guys a little something for their trouble." He flashed his quick smile. "One of the Gs gave me a bottle of what is supposed to be some really good wine. Those guys working the lift in the plane will love it." As we found out at the end of the exercise, they surely did.

WE RECEIVED OUR resupply of food and other items. All was going great. We had only a few training days left, during which we would conduct weapons and demolitions training. To do this we had to receive a resupply packed by our CIA friends. This would occur at night and under semi-tactical conditions. The bundle would contain twenty brand new 9mmUZIs, several thousand rounds of ammo, along with C4 and det cord. But no blasting caps. My engineer, SSG Raybon, had jumped in with all the blasting caps he would need.

The night of the drop, the DZ party (consisting of me, four of my men, and five Gs to help carry the load) moved to the DZ. We set up an inverted L using our flashlights. As the sound of the aircraft grew louder in a clear night dotted with billions of stars, I flashed the code–three long and two short bursts. We could barely make out the MC-130's lumbering body pass over head.

Nothing came out.

We waited.

Still nothing.

The roar of the four engines grew distant. What was going on? All was set up correctly. I'd flashed the correct code. The DZ was a good 300 meters long and lay near the edge of the military reservation where we were training. I got everyone up and on a line with a good twenty meters between us. We walked along the flight path, hoping to find that the MC-130 had dropped late. We covered the entire DZ and even went a couple of hundred meters into the woods at the north end. *Nada.*

There was nothing to do but return to the base camp. Obviously, something had gone wrong with the drop. We'd have to radio back to headquarters and reschedule for the next night. But this didn't seem quite right. I worried about it all the way back.

When we closed in on the base camp, I walked up to Donahue and told him what had happened. He's received more air drops than all of the rest of us put together. The look on his face told me that he wasn't buying the theory that the Air Force hadn't made the drop. This was too highly visible of an operation for that to happen.

Donahue turned to SFC Finney, our senior weapons man, and said, "Finney, take a couple of guys with you and walk the track of the aircraft for at least two miles past the end of the DZ. If some Yahoos find that bundle, we'll have a major international incident on our hands."

Finney grabbed SSG Raybon and SSG Waite, the Team's medic, and took off. By now it was getting close to light. About 1000 hours that morning, Raybon walked into the base camp and told us that they had located the bundle about a mile off the DZ and near a small French village. Thank you, Jesus!

I immediately sent out a recovery party with Raybon in the lead. Early in the afternoon they returned with the weapons, ammo, and demolitions. We commenced training like nothing had happened.

The operation came to a close, and we had a party with all the Gs, packed up our equipment along with the UZIs, and loaded onto a C-130 for our return trip to Bad Tolz. We all felt good about what we'd accomplished. Nothing was ever said about the drop that almost went terribly wrong. Who needed an investigation with all the finger pointing? Not us.

A FEW MONTHS after we returned, I received a letter of congratulations from the Ministry of Defense, Republic of France.

LTC Hutton indorsed the letter to me. The second paragraph of his indorsement read:

```
2. The classification of SPATULE prevents me from
detailing your contribution beyond saying it was
extremely sensitive, important, and significant. In
particular, I congratulate you on the outstanding
quality of instruction provided.
```

SHORTLY AFTER RECEIVING the letter, I found out that I'd be turning the Team over to another captain who needed a command. I'd been in the job for about a year, and it was time. Besides, I wanted to command the Headquarters Company. This job would serve to put an X in the box for Company Command, something every captain had to have if he wanted to make it to Major. I felt sure that I could get command of the company. Not so fast. Hutton had other plans for me first. I was to become the Battalion's S1. Well, crap.

I wrestled with my rucksack while falling at 120 miles per hour from 15,000 feet. My nose and mouth covered with an oxygen mask.

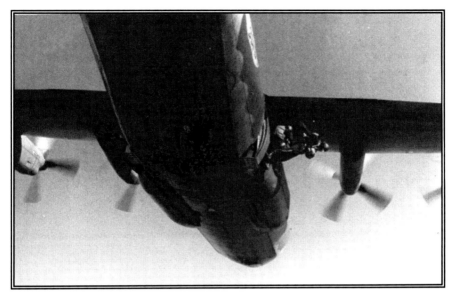

We would exit the aircraft from the side door or off the ramp.

About twenty five jumps later, the course ended. Certificate in hand, I found myself HALO qualified and the Team Leader of ODA 3.

When we got to Augsburg and met our HALO students, standing in the group was my old swim buddy, Preven Jorgensen.

An MC-130E snags the line with a V-shaped yoke mounted on the plane's nose and the individual is reeled on board.

Battalion S1

NOW WOULD BE the time for me to explain to you, dear reader, just how the Army labels its different staff sections. I will be referring to *S* this and *S* that as well as other letters with an assigned number from here on in the book, so. . ..

At the Brigade/Group level and below for pure Army units, S1 refers to a person or group of people who handle all the personnel actions for the unit. S2 refers to a person or group of people who handle all the intelligence actions for the unit. S3 refers to a person or group of people who handle all the plans and operations actions for the unit. S4 refers to a person or group of people who handle all the supply and logistics including maintenance and transportation actions for the unit.

Staff levels above the Brigade/Group are assigned the letter G. The G1 through G4 deal with all the things that Ss do but at a higher level. Normally, G staffs have a commander who is a General Officer.

At the Army Installation level or Community/Kaserne level, you have Directors. For instance the DPCA or Director of Personnel and Community Activities works all the personnel issues for that Army Installation or Community as in Germany.

If the unit consists of personnel from the different services (Army, Navy, Air Force, and/or Marines), the headquarters is referred to as a Joint headquarters, and its staff is designated by a J. The Js perform all the same functions as the Ss and Gs do. This staff is called a Joint Staff.

Now, you might think I'd be through, but you'd be wrong. If the military command consists of members from different countries, like Britain, Turkey, France, and/or the United States, for example, the staff sections are designated as C1 through C4. This staff is called a Combined Staff.

Through yet? Hardly. If the headquarters is both a Joint and Combined staff, the designation becomes CJ1 through CJ4.

There are higher numbered Ss, Gs, Cs, Js, and CJs, but I won't be referring to them in this book, so who cares? You may be asking who cares about the last seven paragraphs. Well, inquiring minds want to know. That, added to the fact that I served on or within S1, G1, J3, and CJ3 level staffs, make it pertinent to the rest of this story. So there.

I HAD WANTED to command the Headquarters Company, but Hutton wanted me as his S1. I'd try for that job later. The Army had gone to the

PAC system. The Personnel Administration Center consolidated all the company clerks under one officer and NCO. This concept was not initially popular with the company commanders, but soon they realized that not having to worry about personnel matters when they went through an IG (Inspector General) inspection was a BIG plus.

SFC Laurence Yeager served as my NCOIC and took charge of all the clerks. I mainly ran interference for him between the majors who commanded the three companies and the Group S1.

I learned to read and interpret Army regulations. No small task. I had to make sure that all admin suspenses were met by all in the battalion, from submitting timely OERs and NCOERs to all the various reports required by the Group S1 types, including the few requirements passed down by the DPCA.

SHORTLY AFTER I took over in the Battalion S1 shop, the Group changed command. COL Wereszynski was replaced by COL John R. Martina, Jr.. Martina was an Artillery officer who had spent time in SF as a captain assigned to the 46th Company in Thailand. He was a short, stocky, gruff man with sandy colored hair and a rough and tumble, take-no-prisoners attitude. Because of this, most officers were a little afraid of him. He was married to Mary, his second wife. Most SF guys were on their second, third, or fourth, wives. You get the picture. Anyway, Mary hailed from Florida and was without a doubt one of the nicest, kindest, most considerate southern ladies I'd ever had the privilege to know. I guess opposites do attract. Actually, once I got to know Martina, I saw that beneath all the growling he had a great sense of humor and a very good heart.

Martina took over like a storm. Both the unit and the community had gone through a series of problems, not the least of which was the Community failing their VII Army IG inspection.

The Group commander also served as the Community Commander in Bad Tolz. To assist him in this, Martina had a deputy Community Commander who was an ADA (Army Air Defense) LTC named Zimmerman. He was a good guy but was in way over his head when it came to dealing with all the problems inherent in running an Army Kaserne.

When the IG results hit his desk, Martina summoned me to his office. As usual he bypassed LTC Hutton, my boss. When I walked into his office, I saw LTC Zimmerman squirming uncomfortably in a chair directly in front of Martina.

"Davis, Zimmerman and his crew over on the Community side screwed up and failed their IG." He shoved a pile of paper across his desk in my direction. "We're going to have a relook in two months. I want you to go in there as the DPCA and fix every deficiency in this report!"

"And you," he pointed his stubby cigar at Zimmerman, "will do what ever Captain Davis tells you to do to help him fix it. Questions?"

Zimmerman had none and the "discussion" ended.

I picked up the IG packet and followed Zimmerman out of Martina's office and down the hall to his office.

I told Zimmerman not to worry, and that we'd get things back together. He almost kissed me.

FOR THE NEXT two months Yeager and I bounced back and forth between our office in the battalion and the admin center at the Community. It was a total mess. Apparently, the NCO in charge had never been checked on by anybody. It wasn't that the guy was incompetent; he was just overwhelmed and didn't know who to go to for help. Anyway, when the relook came around, we passed.

Martina called me back into his office and said, "Okay. You did good." Then waved me out. I hesitated. "Whatta ya want, a medal or something?"

"No. What I want is out of the S1 and to command the Headquarters Company." What did I have to lose.

Martina smiled, shook his head, and said, "Get the hell out of my office."

Well, I hadn't been the S1 but five months and two of them were spent as the S1 and DPCA for the community. My time would come.

WE HAD BEEN living in the beehive for about six months. My great aunt and little fifteen-year-old brother, John, had come for a visit shortly after we got settled in. That summer Mother and my sister Marrlee were scheduled to come for a visit, but at the last minute something happened, and they couldn't make it. John decided that instead of working all summer surveying cotton, he'd use one of their tickets and come see us. He was now sixteen and well traveled, spending time in Spain and Scotland and having made the visit to see us the year before.

He began his odyssey in Atlanta, Georgia, not knowing if he would be able to use the plane ticket from New York to Germany until he arrived in New York. What were Mom and Dad thinking? We expected him to call us to come pick him up when he arrived in Munich. As it happened, he managed to get a flight from New York to Switzerland. From there he caught the train to Munich, then boarded a bus from Munich to Bad Tolz. Polly and I hadn't heard from him since he left Atlanta. At 10 p.m. there came a knock on our door. When I opened it, here stood John. I was amazed.

Over the next few weeks, I introduced John to the NCO club and over flowing German beer steins. Someone had to take responsibility for educating him in the finer things of life. As always, he held his own.

One day while John was still with us, Polly found out that a family who had recently arrived was being assigned quarters on the Kaserne. The housing list by regulation placed you on the list based on how long your name had been there. It turned out that the German who was managing the list would occasionally let someone in ahead of others based on who that person knew. Well, that didn't sit too well with Polly. The next thing I knew, she had stormed into Martina's office and was giving him hell. Well, no one gave Martina hell and got away with it. Except Polly. The one thing that I had learned about Martina was that he appreciated a straight forward approach and had little use for people he could intimidate.

A week later John was helping Polly and me pack to move into a really great house on the Kaserne. It had three bedrooms and a maid's quarters on the second floor. Outside our front door, the patio looked directly into the Alps. We had finally arrived.

SURPRISINGLY MARTINA WAS big on doing things for the dependants. Probably because he was on his second marriage and knew the strain the Army and especially Special Forces put on the family. As a result, during the winter months, he tasked the Battalion to run ski training for any of the dependants who wanted to learn to ski. 10th Group in Bad Tolz had some of the most highly qualified ski instructors in the Army. It was a fact. As a result, Polly and any other of the wives who wanted to take advantage of this opportunity learned to ski and to do so well.

Bad Tolz had one of the only indoor pools in the area. I was big into swimming ever since I graduated from the National Aquatic School the summer before my freshman year in college. However, my Water Safety Instructor (WSI) certification had expired. Martina found out about that and called the European Headquarters for the American Red Cross (ARC) and volunteered the pool if they wanted to come down and run a certification course for WSIs. They took him up on his offer, and he insisted that I be one of the students. Throw Brer Rabbit in the briar patch.

The only problem was that Martina expected me to be a student and his main point of contact for the ARC. This, on top of the S1 duties, caused a strain on my management of competing priorities and limited resources.

Mr. Ed Mudgen, the guy who came down to run the course, was head of all water safety for the ARC and even wrote the current Senior Lifesaving Manual. This was old hat for me, and I did extremely well in the WSI course. So much so that Ed asked that I attend the International Aquatic school that he was scheduled to teach in Garmisch. This would allow him to certify me as an Instructor Trainer. This meant that I could then certify dependent wives as Water Safety and Basic Swimming Instructors. Martina immediately saw this as a way to establish a permanent water safety program for the community. Now, through me, he would have a way to do just that. It wasn't long before every kid who wanted to learn to swim was getting swimming lessons by one of several wives I trained as Water Safety Instructors.

One day while slugging my way through a pile of paperwork that seemed to grow like mushrooms on my desk, I got a call that Martina wanted to talk to me. What now? I knocked on his door and heard a gruff, "Get in here."

Martina waved me to a seat. "We're going to host the VII Corps swimming and diving championships. I want you to put together a team and win the damn thing. Think you can do that?"

I had just finished the requirements for the 50-mile swim and fit program and knew several other guys who were excellent swimmers. "Well, we'll need some time set aside for me and the guys to swim. You know, let their commanders know that this is a priority for the next month, and that you want their guys to work for me a couple of hours a day."

"Not a problem. Make a list of who you want, then write me up a note that I can send to their commanders." Martina waved me out the door.

That's just what I did. I especially wanted to get two guys, Hughes and Cook, on board. They were two of the best swimmers I'd ever seen swim, from the butterfly to breast to free-style. I was strong and fast, but a rock compared to them.

By the time it was over, I had seven or so guys who trained with me daily. We had the advantage of having an indoor pool, so most of the guys stayed in great swimming shape year around. When the day finally came, little Flint Kaserne in Bad Tolz blew the rest of the huge installations in VII Corps away, earning 120 points. The next closest installation to us was Augsburg with 72 points. Hughes and Cook racked up most of the points, but I did my part by coming in first in the 100 meter freestyle, second in the 500 meter freestyle, and second on the three meter dive. All in all a great show.

ANOTHER THING THAT Martina did that would benefit us all was to authorize the University of Southern California to establish a satellite Masters program on the Kaserne. Since we lived in Bavaria, a vacation spot in Europe, we got the best professors USC had to offer. They would come and teach two weeks, have one week off, then teach two more weeks. The students received credit for two courses. Of course we had to complete a thesis. Since research material and facilities were scarce, the school was very flexible as to our topics. I picked Characteristics and Methods of Urban Guerilla Warfare. Actually, this worked out great for me. Not only did the thesis work well for USC, but I was able to rework it and use it again in the Command and General Staff Officer's College and then again in the Army War College. When we completed USC's course of study, we had earned a Masters in Education Administration. Hooah!

CAPTAIN JOHN WAYNE SMITH, the Headquarters Company Commander, had come down on orders and would be leaving soon. I saw my chance and took it. I went to my Battalion Commander, LTC Hutton, who was about to leave himself, and asked for his support in getting the company. He agreed, and we went in to see Martina. I had grown to really like Martina and felt sure he would give me the company. He was happy to.

Headquarters and Headquarters Company Commander

THROUGHOUT MY CAREER, I commanded four Operational A Detachments, two companies, two Battalions, a Readiness Group, and a Joint Special Operations Task Force. The Headquarters and Headquarters Company (HHC) is without a doubt the most frustrating and one of the more difficult commands I ever had.

The problem is that, as the HHC Commander, you have all the responsibility and absolutely no authority. The only people that you rate or have direct control over are your XO, First Sergeant, and the company clerk. The company consist of the unit's (Battalion or Group) commander and all of his staff officers, their NCOs, and their privates. Most of the staff officers outrank you. Most of the senior NCOs are senior to your first sergeant. The unit's commander (your boss) is under your "command." Here's the catch. You are responsible for seeing that everyone participates in all the Army's mandatory training to include the semiannual Physical Training Test. You must fill all support taskings with soldiers who work full time for someone else. It's your responsibility to see to it that all regulations are adhered to. You must, when the occasion demands, dole out any non-judicial punishment in the form of company grade article 15s. Most, if not all, of these requirements cause problems for the various staff officers who are trying to support the unit in their functional areas.

To be a good Headquarters Company Commander, an individual must be a consummate politician, be able to see several shades of grey when it comes to your principles, be smart enough to pick the best battles to fight, and know when to give up and move on to the next challenge. I exemplified all of these attributes at the time. NOT.

The only thing that saves the HHC Commander from going postal is that he works directly for the unit's XO or Deputy Commanding Officer (DCO) who *does* have the hammer over all the staff that the HHC Commander finds himself butting heads with on a daily basis. If the HHC Commander gets along with the DCO, his life is bearable. Most of the time.

WHEN MARTINA CAME to Bad Tolz, he brought along with him a LTC named Don Solan. Solan, a tall, willowy man, was not SF qualified but had commanded a Mechanized Infantry Battalion in northern Germany. At that time, a soldier could take the SF course by correspondence then serve in a position during an FTX commensurate to his rank, and the Group Commander could award him an SF qualification or S prefix. This is exactly what Solan did, and the Special Forces was much stronger as a result. Solan and I would serve off and on in SF for years to come. I admired and respected him to the max. Not only had he commanded a mechanized battalion, he had once been an HHC commander. He saved me from myself more than once. Although at the end of my year as HHC Commander, his hair was decidedly greyer.

≫→◆→≪

I WAS FORTUNATE to have a great XO, Lieutenant Dave Steinberg. Dave was a Cornell graduate and a member of their Light Weight Crew. He was also an avid skydiver. I'd put Dave through jumpmaster school when I commanded ODA 3. He proved to be an exceptional organizer who could accomplish any task assigned. We became great friends.

My First Sergeant at the time was a Master Sergeant named Crowley. He had been in the job for a couple of years and knew all the ropes. This was why, when I put the NCOIC of the S4 shop on the manifest for a static line jump, he walked into my office with the jump manifest in hand and said, "Sir, you've got MSG Jackson (not his real name) on the manifest."

I nodded. "Right. I noticed he hadn't made a static line jump in. . . well, forever."

At the time the Group filled a parachute demonstration team that would conduct demonstration jumps when tasked by the S3 in support of change of commands or other community events. The jumpers were all very experienced skydivers and would receive credit for their jumps for pay based on their skydiving activities. This was apparently within the regulations, but it rubbed me the wrong way. Most, if not all, of the other guys who participated in these demonstrations also jumped static line. Why shouldn't Jackson?

Crowley shook his head. "Sir, he doesn't do static line. Only free fall demonstrations. It's just the way it is."

"*Was*," I said. "It's not right that everyone else has to make static line jumps and he gets to skate. Leave him on the manifest for Friday."

Crowley shook his head, smiled, turned, and walked out of my office.

Dave had heard the conversation and poked his head through the door. Dave was one of the ones who jumped on the demonstration team. "Sir, Crowley's right. This isn't a fight we want to have."

"Might not be, but it isn't right. It's one we'll have if we have to." I waved Dave out.

Within the hour, MAJ Brennan, my old company commander who was now the S4, stormed into my office. "Davis what the hell do you think you are doing putting Jackson on the manifest?" He fanned my face with the paper that had the names of Friday's jumpers.

"Sir, your guy hasn't made a static line jump in. . . well, several years. I thought it would be a good thing if he just made *one*. You know. . . set the example for all the soldiers in the S4." I flashed my best "Come on, let's do it" smile.

"That's bullshit. You just wanna stick it to him. I'm going to Solan. We'll see what he has to say."

I followed Brennan down the hallway and into Solan's office. Solan peered over his glasses and knitted his brow as we walked in. He'd already heard about the blow up and was waiting.

Brennan explained how busy Jackson was and how many jumps he made on the demonstration team and how jumping static line would be an insult to him and so on. I had to admit that Jackson had done more than his share of demonstration jumps, all of which couldn't be classified as fun. After all, who wants to free fall from 10,000 feet and try to land inside a quad of buildings between a flag pole and a reviewing stand? He also served as a static jumpmaster for quite a few static line chopper jumps. Maybe I was having second thoughts.

My argument to Solan was that all the soldiers from private up to colonel made static line jumps. After all, it was how we were supposed to go to war should the need ever arise. I made sure that all the privates and junior NCOs in the S4 shop jumped static line every three months or didn't get their jump pay. Shouldn't the most senior NCO in the shop set the example for everyone else?

I could tell Solan didn't like me putting him into the position I had, but he was big on setting the example and told Brennan that Jackson would make the jump on Friday. End of discussion. I had won. Or had I.

Jackson made the jump on Friday and didn't like it one bit. Surprisingly, I wasn't too happy either. During the rest of my time as HHC Commander, I didn't put Jackson on another static line jump.

LTC HUTTON LEFT the command and was replaced by LTC James A. Guest. Guest stood just over six feet. If a New Orleans street artist were to draw a caricature of him, it would show a face with shoe-size ears and a wide grin full of teeth with a gap between the front two. He was never short of good ideas, *most* of which worked. Typical A type personalties, he, Martina, and Solan would often disagree. When those three elephants butted heads, this monkey took to the trees. I enjoyed watching Guest work. This was a good thing since I would serve under him several times for the remainder of my career.

I HAD ALWAYS been on time everywhere, and had a real thing about it. I know you won't believe this, but I went through four years of college and never cut or was late for a single class. In the Army when you had a formation, you reported your soldiers' status by saying, "All present or accounted for." The "or accounted for" part meant that you knew where everyone was: some were on leave, some were TDY, some on sick call, and so on. A soldier might also have overslept and called in. The bottom line was you knew where all your soldiers were. The problem was that the standard thing to say was, "All present or accounted for" even when someone, especially a high ranking NCO or Officer, hadn't shown up for the formation. I couldn't and wouldn't live with that.

At one of our weekly formations, First Sergeant Crowley wasn't there. I told my XO to check with the company clerk, who I had sitting by the phone in case anyone called to say they would be late. Crowley hadn't called in.

Dave told me that, so when the time came for me to report, I said, "One man absent."

Well, apparently no one had ever reported anyone absent before. Normally, they would cover for the individual and after the formation try to find the guy or guys who were missing then chew them out. When I did otherwise, you'd of thought the world had come to an end. After the formation, the S1, CPT Frank Tabella, marched up to me and asked who was absent. I told him my first sergeant. The next think I knew, Martina had me in his office with Solan, lecturing me on why it was important for me to cover for my first sergeant.

I told them both that if I had reported otherwise, it would've been a lie. Crowley had come in and had no problem with being reported absent. He had been. I'd told him not to do it again, and he agreed. Hey, it happens. Now all the privates in the company knew we meant business when it came to being on time for formations. They definitely wouldn't be late. So there.

Martina fumed and Solan rolled his eyes. They grudgingly knew I was right. But was I really?

ONE OF MY duties as the HHC Commander was to serve as the meet director for the Group's annual international skydiving competition. Sport parachute teams from numerous foreign countries and the Golden Knights from the States attended the event. It was a big deal.

Flintlock was just around the corner, but first we had to get through the skydiving competitions. Dave and I rooted around the office trying to find an after-action report or something we could use as a guideline.

All we found was a few papers in a folder with the addresses of all the countries that were invited. Not much there.

I called Donahue over from the HALO Team. Sure enough, he had some notes and would help. Actually, he would act as chief judge for the competitions since he had more credibility in the skydiving world than anyone around.

We pulled off the event, hosting thirty-two teams from over twenty countries. By the time we finished, Dave had constructed an after action-report that filled two large three ring binders. Whoever got saddled with this bronco next year would start out light years ahead of where we did.

OVER THE PAST seven years, I'd been on five Flintlock exercises, but all of them as an ODA Commander. This exercise I'd be serving as the Headquarters Commandant, staying at the SFOB headquarters in Greenham Common, England. Here I'd tend to all the housekeeping duties for our staff and the ODAs that would pass through on their way to operational areas throughout Europe.

Housekeeping duties included feeding and housing the several hundred soldiers that stayed or passed through our headquarters. It was like running a large hotel with a restaurant. I was mainly responsible for those SF guys who came from Bad Tolz, which consisted of the battalion headquarters along with the companies and ODAs that would pass through day and night. In addition to those individuals, we also had all of the folks from Devens along with their Group Headquarters that would really be running the show.

My old buddy CPT Tony Salandro flew in from Devens to assume the duties for the Headquarters Commandant for the 10th Group. We teamed up and worked together as the job required 24/7 attention. Tony and I had served together off an on for several years. He was married to Fay, a full-blooded Indian, whom he claimed to be deathly afraid of.

One of the childish games we all played was the Coin Challenge. Every SF guy had to keep a Group coin on himself at all times. Normally, the coin was engraved with the soldier's name and unit. The one I currently carried had my name and Cdr ODA 3 on it. While sitting around a bar drinking, someone would slap his coin down–the challenge. Everyone else had to slap his down as well. Those who could not produce a coin had to buy everyone a round of drinks, most often beer. If everyone had their coin, the guy who made the challenge had to buy.

Salandro always carried his coin and was the worst I'd ever seen at "Coining" others. He'd even coin me while on a run. One day I had the bright idea to coin him while he was taking a shower. I was sure I'd get him. Someone must have warned him as when I dropped my coin in the shower, he squatted and his coin fell out from between his butt cheeks. Damn!

This coining challenge thing was getting expensive, and I couldn't figure out how to get the best of Salandro. Then it hit me. I called back to Bad Tolz and told Dave to send over a 1st Battalion coin with Salandro's name engraved on it. I'd wanted to give Tony a coin before we left England anyway.

The coin arrived with the next team, and the Team Leader passed it on to me. I waited until the Club was full. I'd made sure that several of the guys whom Salandro had caught without their coins got the word, so the place was packed.

I slapped the coin I had engraved with Salandro's name on it down on the bar. It started raining coins. When Salandro reached back for his billfold to get his coin, I said, "Salandro, you already got a coin down. In fact it was the first one out."

He started to protest.

"Here." I pushed the coin to him. He picked up and saw his name. "I can't believe you did this!" Actually, he was very pleased.

I WOULD OFTEN sit in on the Teams' brief backs and check with them periodically throughout the time they were in isolation to make sure that all their housekeeping needs were being met.

I remember visiting one Team who had been given a real bear of a mission. It reminded me of several operations I'd gone on in the past. After talking with the Team Leader and Team Sergeant about what they had coming up, I said, "Well, I'm damn glad it's you guys going in instead of me."

The Team Sergeant paused, smiled, then said, "Sir, I've been waiting twenty years to hear some staff officer say something like that instead of the ususal 'Boy, I sure wish I was going in with you guys.'"

I knew the feeling as I had heard that BS from staff officers many times in the past. He appreciated my honesty. And I really was glad it was them and not me–this time.

FLINTLOCK DREW TO a close and we redeployed back to Bad Tolz. I received orders sending me back to the States and to Fort Polk, Louisiana, home of the 5th Mechanized Infantry Division. If I ever planned to make it twenty years in this man's Army, I had to get qualified as an Infantry officer. Where better to test my mettle than in the Mechanized Infantry? Hell, I had trouble spelling "mechanized" and all of a sudden I was going to be one.

After receiving my annual efficiency report as the HHC Commander, I reapplied for a Regular Army Commission. All my bosses gave me great letters of recommendation, especially my old boss LTC Hutton. This time DA approved it, giving me a glimmer of hope that I might succeed in the Army.

Polly and I had begun packing for the return trip to the States when I got a call from the S3 shop. It seemed as they had gotten a request directly from the American Red Cross headquarters in Europe asking if Captain Tom Davis could be sent to Incirlik, Turkey, to run an aquatic school for Air Force personnel. The purpose was to qualify Water Safety Instructors who would then return to their air bases and set up swim programs. The ARC would fund the trip if the unit could spare the Captain.

I really hated that I would have to deploy to Turkey and teach Water Safety Instructors rather than help Polly pack up everything then clean and clear quarters. But. . ..

Polly was not amused. She gave me her "Here we go again" look. But what could I do? They needed me. I was the only one in Europe who could pull a course like this together in four days, deploy to Turkey, and teach it.

So off I went for two weeks of teaching swimming instructors and sitting around drinking beer in the Air Force's Officer's Club. One day I'd return to Incirlik, but it would be under different and more serious circumstances.

Polly, ever the trooper, handled all the packing and cleaning and clearing of quarters. I got back just in time to stuff my clothes in a bag and get on the plane with her and the kids.

One bright spot with leaving the Headquarters Company was that I'd be turning over command to my good friend Zak Kozak. Even though he sported a big mustache, I knew I was leaving the company in good hands.

One bright spot with leaving the Headquarters Company was that I'd be turning over command to my good friend Zak Kozak. Even though he was sporting a big mustache, I knew I was leaving the company in good hands. From left to right: Me, Zak Kozak, and Company First Sergeant Zapata.

Chapter 6

Fort Polk, Louisiana

Combat Support Company 3rd Battalion 10th Infantry

POLLY, THE KIDS, and I pulled into Fort Polk in the middle of the summer. Coming from Bavaria, we were in for a shock. I'd be dripping wet just sitting and not moving. It would take several months to acclimatize to this blistering heat and stifling humidity. Leesville, Louisiana, was a typical Army town. Polly wanted little to do with it, so we started house hunting in nearby (read that 35 miles away) Deridder. We found a very nice (by my standards) three bedroom house and bought if for $42,000. Of course, this meant that I'd have to get up every morning around 0430 to make it on post in time for PT.

I SIGNED IN off leave and reported into the 5th Mechanized Infantry Division's G1's office. To my front, behind a large mahogany desk, sat LTC Scott, a thin, wiry 50-year-old with salt-and-pepper hair. He pulled his military issue black horned rimmed glasses off, placed them on the desk, and looked up from reading my ORB (Officer Records Brief). "Captain Davis, are you sure you want to command a company here in the 5th ID? I see where you've already had a Headquarters Company."

Even though I had commanded four ODAs and a Headquarter Company, Infantry branch didn't consider me "branch qualified," thus a promotion risk to Major, so I answered him, "Yes, sir. I'd like the chance to command a mech company if at all possible."

LTC Scott shook his head. "You know it doesn't matter how well you have commanded in the past; if you screw up in an additional command you'll never recover from it."

I nodded that I knew. In fact, my old buddy Spurgeon Ambrose had been preaching this to me that for years. "Sir, I don't plan on screwing anything up, and I'd like a shot at another command."

The G1 sent me to the 2nd Brigade who passed me down to the 3rd of the 10th. I reported to the battalion commander, LTC Jerry Webb, a tall man with light brown hair and a reddish complexion, sporting an

Americal Division combat patch on one shoulder and the 5th ID's red diamond on the other.

Webb welcomed me to the battalion and told me that he was giving me the Combat Support Company (CSC). He went on to explain that the Division was experimenting with placing all the battalion's TOWs (a heavy antitank weapon) into one company and that they would be in *my* company. This would give me 36 tracks and 13 wheeled vehicles, making the company the largest in the division. Additionally, the company had within it the Scout Platoon, Air Defense and Ground Radar sections, and the Heavy Weapons (4.2MM Mortar) Platoon. All of these supported the Battalion when it went into combat or on Field Training Exercises.

I thanked him, walked down to the company, and began my change of command inventory. Three weeks later and with a stack of 3161s documenting about 2000 items the company was short, I took command.

Right after the change of command, I sat down with my 1st Sergeant, a short, light-skinned Mexican named Serna. First Sergeant Serna wore his dark hair cut high-and-tight and peered at me through black horn-rimmed glasses. What hair he left on the top of his head, he combed straight down across his forehead. He spoke with a slight accent as he briefed me on all the NCOs in the battalion and only when I asked did he give me a rundown on the officers.

I next gathered my XO, the Platoon Leaders, their Platoon Sergeants, and Serna into my office and introduced myself. I told them generally what I expected of them and explained that the fact that I wore a Special Forces Combat patch didn't mean that I was not happy to be in the mechanized infantry (and I meant it). I also told them that after almost ten years in the Army, this was my first tour with the mechanized infantry and that what I knew about maintenance you could put in a very small thimble and that I'd need their help in this area.

THE COMPANY HAD a PT formation every day as 0600. After calisthenics, the company would run as a group at a painfully slow pace. It was killing me, so I decided that I would break the company down into three groups: the eagles, the owls, and the grim reapers. Fast, medium, and slow. I led the eagles on their runs. Once a week I'd take them on the "adventure trail." We'd run out of the company area and down to the rifle range. Here I'd lead them up and down the thirty-foot embankment that served as a back drop to the range. From there we'd hit a small creek that varied in depth from ankle to knee deep. After about a half mile, we'd cross underneath the road via a large culvert then return to the company area. Good times were had by all. Or at least by me.

SHORTLY AFTER I got into the Division, I ran into Bob Glandville. Bob and I served together at Devens. He was in the other Battalion there, and I didn't know him really well. Bob was medium height with a little bit of a gut, but his legs were like tree stumps. He could run all day. It was he who got me training for the New Orleans' Mardi Gras Marathon. This marathon was greatly favored by runners who wanted to qualify for Boston. The course crossed the Lake Pontchartrain bridge, the longest bridge in the US. It was totally flat. Race officials had the ability to start the race at whichever end that would place the wind to the runners' backs.

Of course on the day Bob and I ran the race, they screwed up and started the race with runners running into the wind. It took me well over four hours and Bob just under four to make it across the finish line. They were hauling disappointed and tired runners off the course by the truck load.

Polly and Diane, Bob's wife, waited for us at the finish line, with impatient looks on their faces. Where had we been all that time?

Bob was also the guy who got me started with Triathlons, an event that combined swimming, biking, and running. I was a very good long-distance runner and swimmer, so the Triathlon was made for me. I had been running and swimming a lot, but didn't own a bicycle. What could be hard about peddling a bike?

The first Triathlon I did, I did with Bob in Philadelphia, Mississippi. I'd borrowed a bicycle from a friend for the race. That area of Mississippi was washboard hilly. I was one of the first ones out of the one-mile swim, and hit the bike for the twenty-mile ride. All was going great until I got off the bike to start the 10K run. There I met Jesus, Buddha, Muhammed, a couple of dead Popes, and several other religious icons. I wasn't in any way prepared for pain involved in the transition from the bike to the run. Something I'd never let happen to me again. This, dear reader, began my love affair with Triathlons. When my knees and ankles finally gave out on me years later, I'd competed in well over 100 of these multi-sport and ultra-endurance events, often winning or placing in my age group.

Bob and Diane bought a house only a couple of miles from ours in Deridder. I bought a bass/ski boat and with us, Bob, Diane, their son Lee who was Tee's age, and their daughter Jennifer spent many a happy day on Toledo Bend Reservoir. A few years later we'd all serve together at Fort Bragg, and they would buy a place only two houses down from ours.

WHENEVER I had any spare time, I would walk down to the motor pool, check in with SFC Fazzino, my Motor Sergeant, then find the least busy mechanic and ask questions about maintenance: how to do what to this and that, things to check to see if all maintenance was being conducted to standards, and so on.

I noticed that after several weeks of this, my mechanics seemed uneasy with my never ending questions and that Fazzino grew more and more uncomfortable when I dropped by.

I was well into my 3rd month in command and sitting at my desk behind a mountain of paperwork when I looked up to see Serna standing in the doorway with Fazzino behind him. "Sir, Fazzino has something he wants to talk to you about. Can we come in?"

I leaned back and waved them in. Serna sat in a chair to my right and Fazzino took the one directly to my front. Fazzino sat looking down, grinding his right thumb into the palm of his left hand.

Serna leaned forward. "OK, Fazzino you got something to say to the Old Man. Say it."

Beads of sweat popped out on Fazzino's forehead. "Sir, I want to leave the company."

Fazzino was, without a doubt, the best Motor Sergeant in the battalion if not the brigade. I had come to rely on him to keep the massive maintenance headache I had on a daily basis to a minimum throb.

"What's the problem," I said.

Fazzino hesitated a moment, and Serna jumped in. "Tell him what you told me."

"Sir, I can't work for you." Fazzino shook his head.

I was prepared for a lot of things, but this wasn't one of them. "What's the problem."

Fazzino squirmed in his chair. "Sir, it's obvious to me and my mechanics that you don't trust us. I just can't work for you. Look at me, I'm sweating my ass off just sitting here."

After commanding four ODAs and spending over seven years in the trenches with Special Forces, one thing I was, was comfortable around NCOs. I couldn't believe that Fazzino felt this way about me. "What exactly is it that I've done to make you feel this way?" I had to know.

"Sir, you come down to the motor pool every day and ask us all these dumb questions that we all know you know the answer to. After all, no one could be that ignorant about this stuff. It's obvious that you don't think we are doing our jobs. That's it."

Serna smiled, shook his head, and rolled his eyes. "Look Fazzino, when it comes to maintenance, the Old Man is absolutely that ignorant. He basically doesn't know his ass from his elbow."

"Top's right," I said. "I really am that ignorant. Look. I didn't want you to have to set up some formal training program for me. You don't have time to be doing that. I was just trying to learn all I could about something I absolutely have to know about if I'm going to make it, commanding a mech company the size of this one."

A smile broke across Fazzino's face, and I continued, "Not only do I like and respect you, since you're the best motor sergeant in the brigade, I have to have you to keep my ass out of the Battalion Commander's office. I'd really appreciate it if you would stick with us."

Fazzino looked as if a giant boulder had been lifted out of his rucksack. He and Serna stood. Serna thumbed toward the door. "OK, Fazzino, get your ass outta here and back to the motor pool. And no more whining!"

WITHIN A MONTH of taking command of CSC, we went to the field for training with the Battalion. After the first night, I was disappointed to find that the area where we had set up was littered with empty boxes from C rations along with empty C ration cans, paper wrappers, cigarette butts, and, last but not least, used toilet paper. To say that I was royally pissed would be a gross understatement. Later I found out that this was considered standard and that the units would just come back after the field training and police the area. Wrong answer.

The next day I rounded up all the Platoon Leaders and their Platoon Sergeants and commenced to chew on their asses. "I can't believe all the cigarette butts I found in the area where we set up last night. It's disgusting and disgraceful and unacceptable." I jabbed my finger at the group. "Not only does this show a lack of discipline, it can provide the enemy with valuable intelligence as to the size and location of our unit."

I stormed back and forth in front of the group. "And furthermore, what's with all the toilet paper and crap I see all outside the parameter? Anyone want to explain that?" The men were getting the message loud and clear. I was careful not to mention anything about all the other trash, rather throwing my fit about something as small as cigarette butts and a little toilet paper.

"I want to show you guys something." I pulled out my entrenching tool and cut a small square in the grass. Carefully removing the square I set it aside. I then dug a few more shovels full of dirt out. "Okay, when you or your guys want to take a dump this is what you'll do. Bite your loaf right here in the hole. When you're done, drop your used T-paper in, then fill up the hole. Lastly, place this little square of grass over the hole and tap it down." I demonstrated filling the hole back in and replacing the square of grass. "This'll make it harder for the enemy to know we were here. Got any questions?"

All heads shook "no" at the same time. Their eyes wide, I think they all thought I was about to have a stroke. How could something as small as cigarette butts and toilet paper get the Old Man so pissed off?

I finished by telling all of them that if I found any more butts or T-paper around their area, I'd give them a letter of reprimand and if that didn't work, I'd fire them and their Squad Leaders. Now go pass the word on.

Well, after that, I still found the occasional butt around but never did I see any other trash. I guess they all thought that if I would go that postal over something as small as a cigarette butt, no one wanted to see how I'd react to garbage strung all over the area. Which was exactly my point.

ONE DAY A tasking came down to the company. We would be the test unit for the M109 ITV (Improved TOW Vehicle). This weapon system would replace the current mechanized TOW vehicle. It had enhanced night vision capabilities and afforded the gunner the ability to fire the wire guided TOW missile from within the vehicles itself, never exposing him to small arms fire. It was a big deal.

Guys from both the TOW Platoon and Scout Platoon went through three weeks of training conducted by instructors from Ft Benning, Georgia. More than 50% of the training was conducted at night. This meant that the guys were constantly on the move, putting maximum stress on them. At the end of the test, they conducted live fire, during which they fired over fifty missiles. All performed extremely well. I was proud of them, and they were proud of themselves.

THE BRIGADE HAD a change of command, and the new commander, COL Mo Oliver had taken over. Oliver was known to be a real terror, and many of the officers were running a little scared. As part of his "getting to know the Brigade," his S1 scheduled him to visit each company and receive a briefing by the Company Commanders.

I asked Serna to step into my office. "Top, I'm thinking that when the Brigade Commander comes by for his visit that I'd like to turn him over to you and the platoon sergeants to get his briefing."

Serna pushed out his lower lip and slowly nodded "yes."

"Okay, let's get the platoon leaders, their platoon sergeants, Fazzino, and Baily in here. I'll tell them what we want."

At the meeting, I told everyone that the NCOs would escort Oliver around. The platoon leaders were to stand in the background or better yet leave the area. Several of them squirmed in their chairs. Let the NCOs brief the Brigade Commander? This was *never* done.

When the day came, LTC Webb, my Battalion Commander, and a small entourage entered the company's orderly room. Serna called attention, and I walked out of my office and up to Oliver and crew. "Sir, Captain Davis. Welcome to CSC. This is First Sergeant Serna." I waved toward Serna. "He'll be escorting you through the company area. Any questions you have, he'll answer." And with that I turned and walked back into my office. Webb was startled and kept looking back in my direction as Serna led the crew off toward our supply room, where SGT Bailey, my supply sergeant, was nervously pacing back and forth, trying to control his breathing.

When the tour finished, Serna brought all back to the orderly room where I was waiting. "Sir, were all your question answered satisfactorily?"

Oliver nodded and said, "Very much so." He then smiled and all cleared out of the place. I got back to work.

A few months later at a Brigade Hail and Farewell, Oliver pulled me off to the side and said, "You know, when I came around back then, yours was the only company in the Brigade whose NCOs ran the show. I really liked that."

I knew he would.

I FOUND MYSELF well into my command when the 2nd Brigade, along with my battalion, deployed to the National Training Center at Fort Irwin, California. Irwin is a sprawling training center that sits in the middle of the desert. The deployment lasted seven weeks. My platoons were spread out across the desert floor in support of the Battalion's operations. All acquitted themselves well and received high praise from both the Brigade and Battalion commanders.

When we finally returned to Fort Polk, I remember stepping off the aircraft that ferried us back to Barksdale AFB and being stunned by the green surrounding us. I'd taken for granted the lush trees and grass that covered Louisiana. Almost two months in the desert gave me a great appreciation for what we enjoyed in this most beautiful state. I'd not take it for granted ever again.

TIME MARCHED ON, and a year had passed since I took command of the company. The battalion's annual IG inspection loomed only one week away. As usual, I found myself sitting behind a mound of paperwork on my disk. I looked up to see First Sergeant Serna standing in the doorway with SFC Fazzino hiding behind him. I instantly knew from the look on the Serna's face that I was about to become a very unhappy camper. "Sir, Fazzino's got some bad news." Serna stepped in with Fazzino trailing close behind. "Tell him, Fazzino."

"Sir, there's no good way to put this. We are missing the first month of this year's document register."

Missing part of the maintenance section's document register was like the armorer coming up short a weapon or two, and the IG was barely a week away.

"What the hell happened?" I looked at Serna, then at Fazzino.

"Tell him." Serna nodded at Fazinno.

"Well, sir, it's like this. . . .." Fazzino explained what happened as I sat there barely able to contain myself. When he finished I said, "Is that the truth? Who do you think's going to believe that?"

"Sir, I swear it's true."

I looked at Serna. He nodded.

Being a good Special Forces soldier, I assessed the situation and knew that there was only one hope to save myself a major kick in the balls. "OK, Fazzino. Go down to the motor pool and round up the mechanics that know what happened and bring 'em to my office."

Fifteen minutes later, Fazzino, Serna, and three mechanics crowded in front of me. I gave the mechanics the *Readers Digest* version of what

Fazzino had told me and asked them if this was true. They all gave me the Billy Goat. I turned to Serna and said, "Top, take 'em all outside and have 'em write up their statements, get the clerk to type 'em up, and have the XO swear 'em to it. Bring the statements to me. I'll write a Memorandum for Record on this, and we'll put it with the document register. Let the chips fall where they may."

Three days into the IG inspection, Serna stuck his head into my office. "Fazzino just called from the motor pool. You'd better get down there ASAP!"

I beat boots to the motor pool and headed directly for Fazzino's office where the IG inspector sat with the document register spread out on the table.

The inspector, an old master sergeant, looked as if he had pulled maintenance on General Patton's tank. He sat behind a field table we'd set up for him. Tears streamed down his face as howls of laughter rolled from his chest. He couldn't talk. All he could do was shake his head and point at my MFR with its attachments that read:

```
4 Aug 1970
SUBJECT: Memorandum for Record, Maintenance Records

1. Reference attached statements dated 3 Aug 1979.

2. During one night in early January, 1979, after duty
hours, the Document Register for the Combat Support
Company was destroyed. At that time, the Motor Pool
Records for this company were being kept in a large tent
in our temporary Motor Pool. Due to the actions of a
small wild animal, (which is assumed to have been a
raccoon), the records were made completely unreadable
and unserviceable.

3. There was no corrective action possible.

THOMAS H. DAVIS
CPT, INF
COMMANDING

Attached Statement # 1
3 Aug 1979

One evening during the first week of January, 1979, a
raccoon got into our Motor Pool tent and shit all over
the Document Register. The paper work was on my desk and
the Raccoon got all the paper work and most of my desk.
He also got a filing cabinet. We set a trap to try to
catch the raccoon but he didn't take the bait. The
```

raccoon destroyed all the first month of this year's Document Register pages and peeled the paint on the filing cabinet.

SAMUEL M. FAZZINO
SFC E-7
MOTOR SERGEANT

Attached Statement #2

3 Aug 1979

Around about the first week of January, while our Motor Pool was operating out of this big fucking tent, a frisky raccoon took himself a tour of our working facilities. In the process he shit on SFC Fazzino's desk destroying the first part of this year's Document Register. He also shit on top of a filing cabinet.

KEVIN L. GIBBS
SP4 E-4
TRACK MECHANIC

Attached Statement #3
3 Aug 1979

I, Sergeant Howard J. Cook, freely make this statement. Sometime in early January, 1979, while the Motor Pool was working out of a tent, a raccoon got into the tent and shit on SFC Fazzino's desk and a filing cabinet. In doing so, he destroyed a lot of paper work. I built a trap in an attempt to catch him; this however, proved fruitless.

HOWARD J. COOK
SGT E-5
SR RECOVERY NCO

Attached Statement #4
3 Aug 1979

One day in January, 1979, when we were living in a big tent as our Motor Pool office, a Raccoon walked into the tent and shit on a filing cabinet. Then walked over to SFC Fazzino's desk and tore off some numbers on some paper work. The next day SGT Cook tried to catch the Raccoon but the Raccoon was smarter than him.

JAMES K. EZELL
SP4 E-4
TRACK MECHANIC

And that was our story and we stuck to it.

THE INSPECTION ENDED with the IG only finding four minor deficiencies in our maintenance area. No mention was officially made about the missing document register pages. In fact, the IG commended the company for its outstanding performance.

I later found out that the "Racoon Story" had spread through the brigade and even wormed its way up to the Commanding General's office. Big laughs were had by all.

AT THE END of my year in command, the company had successfully deployed to the National Training Center at Fort Irwin, California, completed several field training exercises at Fort Polk, been selected as the test unit for a major Army weapons system, and undergone a very successful IG inspection. And to top this all off, I came out on the promotion list for Major. All in all a right satisfying year there in the Mechanized Infantry.

But all fun things had to end, and as a Captain promotable, I had to leave the company. But where would they put me? Into the G1 as the Reenlistment Officer for the Installation and the Division. I know you're thinking that when I got the word, I wanted to slit my wrist. Well, you'd be wrong.

In the Army at that time, officers had their branch and also an alternate speciality. My alternate speciality was 41 (Personnel Management). I'd been a Battalion S1 and, for a short time, a DPCA. Now, I would strengthen my speciality at the division level serving in a personnel management position. I also saw this as an opportunity to really help both commanders as well as their soldiers. MG Joseph Palastra was the Division Commander, and he was very hot on reenlistment.

Also, little did I know it at the time, but I would do something that would greatly help good soldiers and improve reenlistment Army wide.

We would be the test unit for the M109 ITV (Improved TOW Vehicle).

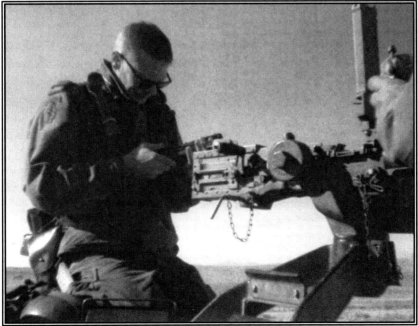

My battalion deployed to the National Training Center at Ft. Irwin, California.

5th Division and Installation's Reenlistment Officer

WHEN I SERVED in Special Forces, I'd never heard of reenlistment interviews. The reg required that Company-level Commanders interview their soldiers on a regular basis. In the conventional Army, the closer the soldiers got to their ETS (Estimated Time of Separation), the more intense the pressure Commanders brought on the young specialist or NCO, explaining how beneficial it would be for him or her to give the old Green Machine another three or four years. The Army offered meager reenlistment bonuses and the option to change jobs or duty stations, but it was a hard sell. During the Carter years, a young sergeant E5 who was married with two kids and whose wife didn't work qualified for food stamps. This wouldn't change until President Regan was elected in 1981.

Now, none of us joined or stayed in the Army to make a fortune, but shouldn't a soldier expect to feed his family without government aid?

Reenlistment was not only a big thing for MG Palastra, Commanding General for the 5th Mechanized Division, it was a top priority for the Army as a whole. In order to sustain the "All Volunteer Army," commanders, recruiters, and those of us in the reenlistment business had to work overtime.

Because of the pressure brought to bear throughout the division by Palastra, commanders at all levels were involved in the effort to retain quality soldiers. If a Company Commander couldn't convince one of his soldiers who did not require a waver to reenlist, the soldier was passed to the Battalion Commander. If that didn't work, the Brigade Commander would take a shot at it. I'd spend many an hour with commanders at all levels trying to enlighten soldiers on the benefits of sticking around for another tour.

The regulation didn't require this level of command involvement. Palastra did. As a result, good old Fort Polk, considered by some as the armpit of the Army, consistently placed number one in reenlistment in all of FORSCOM (U.S. Army Forces Command). FORSCOM had overall command of all the Army's active component major subordinate commands, including several U.S. Armies and Corps.

There was rarely a week that passed that a reenlistment officer from another post or division didn't call, wanting to know how we did so well in reenlistment. I'm sure they all thought we were somehow cooking the books. My answer was always the same: Direct command involvement at every level from Division down to Company. Few if any other Division Commanders were willing to apply the kind of pressure through the ranks that Palastra did. Well, their loss. Our gain.

What made this even more remarkable was the fact that Palastra wouldn't grant any wavers for reenlistment. If a guy or gal had an article 15 or held any status like being barred or flagged that required a waver, they were a no go. Tough standards.

The only area where he showed flexibility was granting wavers to soldiers who were overweight but passed the tape test. Soldiers who tipped the scales based on an arbitrary number of pounds for their height, could be "tape tested" to determine if they had excess body fat. According the Standards of Medical Fitness, body fat was calculated based on the physical size of a person, which includes height/neck/waist size for men, and the height/neck/waist/hip size for women. If the soldiers didn't tape out, they would be placed into a weight-control program. If the soldier taped out, like a weight lifter did for instance, he could reenlist. He still required a waver, and Palastra was okay with this.

This weight thing got started back in the mid 70's when some unnamed Soviet General made a comment to the Army Chief of Staff about how fat and dumpy the American soldier was. As usual, the Army overreacted and laid down the law in an effort to root out all the tubby troopers we had in our midst. I have to say that even though this was a painful process, it did improve the overall fitness of the force. So thank you, unnamed Soviet General.

ONE DAY AS I sat in my cubicle studying the latest Division reenlistment statistics, I heard a heated discussion between my Sergeant Major and one of the Brigade reenlistment NCOs.

"This is total bullshit," said the voice belonging to the 2nd Brigade reenlistment NCO.

"Well that may be, but it's right here in the reg." I heard a finger stabbing paper.

"Look at this guy. He can lift the rear end of a three-quarter ton truck and not break a sweat."

Well, that piqued my attention. I stepped around my cubical. SGM Grover sat behind his desk, his narrow face wrinkled at the brow. In his hand he held the black three-ring binder that housed AR 601-280, the reenlistment regulation.

The Brigade reenlistment NCO stood to Grover's left leaning in to see what Grover pointed at. A young specialist 4th class sat in a grey metal Army chair. His shoulders hung on him like the axle of a deuce and a half. The rolled up sleeves of his fatigues were so tight around his massive biceps that they had to be stopping all blood flow to his catcher's-mitt hands.

I nodded at the SP4. He nodded a melon head sitting on a tree-trunk neck back at me. God, this guy was huge!

"What's all the ruckus about?" I said to Grover.

Surprised, Grover looked up while still pointing at the reg. "The soldier over there," he nodded toward the giant, "wants to reenlist for Benning to be near his parents in Colombus. His mom's just been diagnosed with MS."

"So, what's the problem." I took a seat in another empty grey chair.

"Well, he didn't meet the height and weight requirements and had to be taped." Grover flashed a "That's the problem" look at me.

"I'm still not seeing the problem. Obviously, this guy passed the taping." What was I missing here.

"Well, it says right here in 601-280," Grover stabbed even harder at the reg, "that if a soldier requires a waver he can only reenlist for present duty assignment. In his case," Grover nodded at the giant whose frown was growing deeper with every word, "it means Fort Polk, so he's getting out."

"What? How long has this been around?" I turned in my chair, so I could better face Grover.

"Well, a long time in 601-280. The problem is that this waver using a tape test for overweight soldiers has just recently come out. I'd stumbled on to this the other day when I was researching a way to get around some of the CG's guidance in this area." Grover removed his glasses and sat them down beside the three-ring binder.

"Okay. Let me get this straight. This soldier," I glanced at the giant, "is taped, found with basically zero body fat, and is granted a waver to reenlist, but because he is in such good shape that he qualified for a waver, now can't reenlist except to stay here. Not that staying here is a bad deal," I hastened to add.

Grover and the Brigade reenlistment NCO nodded in unison.

Only in the Army could this catch 22 make sense. In SF I'd never even conducted a reenlistment interview. It was understood that all the NCOs would reenlist, and if they didn't want to be a part of SF any more, well, screw 'em. And so far as weight went, like my old Battalion Commander in Bad Tolz, LTC Guest, said when he heard about the new weight requirements, "Screw it. I'm not going to subject my soldiers to some hair brain idea dreamed up by some jackass staff officer that will be soon forgotten."

Well, Guest had a point but only from the perspective that the vast majority of SF soldiers who didn't pass the weight requirements were built similarly to the giant sitting to my left. As a matter of fact, I never recall ever weighing any of my soldiers, much less taping them.

If we enforced this part of the regulation, it would present a major problem for us here in the armpit of the Army. Not all, but a good number of soldiers, wanted to change MOS or duty station to reenlist. Here we were saying to them, in effect, since you are hard as a rock and keep yourself in such great shape, you're going to require a waver to reenlist, and we can't help you get what you want. Thank you very much.

The wheels in my brain jumped into overdrive. "This is total crap. Get me the number of the proponent for 601-280 in MILPERCEN." Grover rattled it off the top of his head, wrote it down, and handed it to me.

"Wait one," I said, and returned to my tiny cubicle.

As luck would have it, the Major who I finally got on the phone was an Infantry officer doing his penitence in MILPERCEN and not one bit happy about it. "MAJ Dugan," was the crisp answer I heard.

"Dugan, I'm MAJ Davis, reenlistment officer for the 5th Division here at Polk." I could hear Dugan sitting up in his chair. Polk had lead FORSCOM for just about every month for the past year in reenlistment. We were the golden boy when it came to that.

"Hey, you guys must be cheating to make the numbers you do." I could hear a smile in his voice.

"Whatever it takes, my man. Listen. I've gotta problem that I need a little help with from a guy who's been in the trenches. Would that be you?" I was stroking him hard, and it was paying off.

"Oh, God! What I'd give to get out of this asshole town." He meant Washington D. C..

"I hear that. Here's the deal. I've just found out that if a soldier passes the tape test to the tubby trooper reg, he gets a waiver to reenlist."

Dugan broke in with a weary, "Yea. That's right."

"Okay. Here's the rub. I've got a guy who looks like he could tear the turret off a tank and eat it for breakfast. I'm an old SF guy, and we are full of guys like that. You know. The type you want next to you in a foxhole when the bullets start flying. Got it?"

"Yeah." Another weary reply.

"Okay, so 601-280 says—now remember 601-280 has been around long before the Army went postal about tubby troopers—that if a guy has to have a waver, which in essence is what being taped causes, he can only reenlist for present duty assignment." I paused to a long silence on the other end. I could almost hear Dugan's wheels turning as mine had. After all he was an Infantry officer just like me.

"I'm getting the picture." Dugan said. "We're punishing guys who are physically fit. Weight lifters even. Hummm."

I could see the light flash on over his head from all the way down here in Louisiana.

"Okay. Here's what we'll do. You write up a quick request to change 601-280. Say something like if a soldier passes the tape test even though he requires a waver, he can reenlist for any options that are available to a soldier who does not require a waver. Something like that. I need this initiated from the field. I'll massage it and put it into MILPERCENise then walk it through the system." I could hear the excitement cackling in his voice.

"And what about my guy?" By now Grover and the other two guys were standing in front of me with "at-a-boy" looks on their faces. I was on a roll and loving it.

"I'll get my branch chief to fax you a waver letting you reenlist your guy for whatever he is eligible for without a waver. That should do it until we can get a change to the reg out to the field."

I hung up and smiled at the group. "Reenlist him for Benning. We'll be getting an exception to the reg sometime today." High fives were slapped all around.

A week or two later the Army sent out an amendment to AR 601-280 notifying all of the change Dugan and I had negotiated. Needless to say that when the word got around, I was a hero from company level all the way up to the Division CG's office. It really felt good slaying at least one dragon.

I WAS BUMPING up on the end of my third year in the 5th Mech when my office phone rang. I picked it up. "MAJ Davis."

"Hey, Davis. This is your friendly Infantry Officer Assignments guy, Dimsdale calling. Wanna know why?"

Well I knew why and had been expecting a call. Time to saddle up and move out. Hooah! "Let me guess. You've got an offer I can't refuse. Right?"

Dimsdale didn't miss a beat. "How'd you like to go down to sunny Puerto Rico and teach in the University's ROTC Department?"

Well, of all the assignments I'd contemplated, this definitely wasn't one of them. Puerto Rico—sun, surf, scuba diving in crystal clear water, drinking rum on the beach. Duh. What could I say. "Send me in, coach!"

"Great, I'll have orders cut on you, and you can move out in a couple of months. Give you time to take a little leave." Dimsdale was excited. He'd filled a requirement and made a soldier happy at the same time. Something that didn't always happen.

I called Polly to give her the good news. She was cautiously excited.

I've got to call Spurgeon, I thought, then grabbed the phone and dialed his number. MAJ Spurgeon Ambrose was serving his sentence in MILPERCEN in the Officer Distribution Division. He worked closely with all the branch assignment guys. Maybe he had had something to do with me getting such a cushy assignment. He'd be excited for me.

When Spurgeon got on the phone and heard my "good news," there was a grave and ominous silence on the other end. Uh, oh. Something wasn't right.

"Puerto Rico ROTC." The tone of Spurgeon's voice had a "You're screwed" sound to it.

"Yeah. Great. Huh?" I was getting the feeling it wasn't, but one could always hope.

"No. Not great. Bad. Let me tell you something, brother Tom, if you don't speak Spanish down there and I mean fluently," Spurgeon was well aware of my language ability or lack thereof, "you'll get your lunch eaten on a daily basis. You've always done it your way, skating by with all this SF time and somehow managed to survive, but you *won't* survive this one, old friend.

"Oh, crap! I told Dimsdale I'd take the assignment. What'll I do?"

"Sit tight. Will get back to you." And the line went dead. I sat there staring at the phone, feeling like I was going to throw up.

THE NEXT DAY Dimsdale called back and apologized for not knowing the skinny about Puerto Rico and offered me a job teaching in the ROTC department of Auburn University. That Spurgeon. You had to love him.

Later in the week, I got a call from the PMS of Auburn welcoming me to the department. He was happy to get a guy with my background.

I'd done it all in SF and had survived in the mech Infantry. A rare combination indeed.

BUT IT WAS not to be. A miracle happened. I popped on the list to attend the United States Army Command and General Staff College at Fort Leavenworth, Kansas. For those of you who don't know the ins and outs of the Army, at that time, the selection process for this school placed the officer well within the top one third of peers. Not only did it mean that promotion to Lieutenant Colonel was a likely possibility, even for me, it could mean that if I didn't screw up, a long shot for me, and got all the right jobs, I could even be selected to command a battalion. Something I'd dared not let myself hope.

This was really a cause for celebration. I called Polly first and Spurgeon second. For some reason, Spurgeon wasn't surprised. He's gotten a peek at the list and knew, but couldn't tell me. That Spurgeon.

WE PUT OUR house on the market and sold it for $52,000, earning us a $10,000 profit. I stopped by Ron Robert's office. He was with Ed Jones investments. I wanted to discuss what I should with my fortune. He suggested that I put it into a new company with a funny name, Apple. A guy named Jobs, what kind of a name was that, was making computers for individuals. What the hell would a person do with his own computer? I decided to put it all in a Templeton mutual fund. Couldn't sucker in a financial genius like me. Warren Buffett. Eat my dust.

Chapter 7

Fort Leavenworth, Kansas

The Command and General Staff College (CGSC)

POLLY AND I put Fort Polk in our rear window and drove to Leavenworth, Kansas, I in my little dark blue Dodge Omni and Polly with the kids in the light blue eight-passenger van we'd bought in Germany just before we left. We'd had it delivered to a dealer in Macon, Georgia. It sat waiting for us when we got there. The van was pretty plain. After picking it up, we had a white stripe painted around it and, thereafter, called it the Easter Egg. In Deridder, Louisiana, some kids, out for a little fun, had shot the back end out with a shotgun while the Easter Egg sat in our driveway. The Sheriff caught them. Their parents paid me in cash for the repairs. I replaced the back window, then patched the spattering of holes, filling them in with Shoe Goo and covering them with touch-up paint. I pocketed three hundred dollars in the process. Anyway, Polly wasn't all that happy with it, but she, as always, would make do.

When we arrived at Fort Leavenworth, we were assigned housing. It left something to be desired. Our little two-story, three-bedroom place had a basement and was sandwiched in the middle of three other old houses exactly like ours. With two kids, a wife, and a big black and white Newfoundland, I had to admit that we were a little crowded. But, hey, it would only be for nine months. Anyway, it had beds, an indoor toilet, a refrigerator, and a stove. What more did we need? Right?

I FOUND THAT the nine months of hard studying I'd spent in the Infantry Officer's Advanced Course, a change forced on the students by the school's new commandant, BG Richardson, really paid off at CGSC. I had the Tactics and Operations Order classes down pat. The time spent along the way in various personnel assignments writing papers and researching regulations helped hone that skill as well. Before I left Polk, I had written all ten papers that the school had said we would be required to turn in at some point during the course. Unfortunately or fortunately depending your perspective, the school added Physical Training, and thus had to cut some student requirements. Guess which requirements they cut.

At CGSC, we learned all the ins and outs of higher level staff functions, as well as tactics. But by far the best thing I got out of the course was meeting the other officers in attendance. Our forty-person classes were broken down into ten-person groups. Each group had a Group Leader, the senior major. Our "Leader" was a Transportation Officer named Monty Montero, who was of Portuguese extraction. He was married to a great girl, Candy. Monty was much shorter than me and a little pudgy. What did you expect from a Combat Support type?

Sitting next to me was a guy named Tom Futch. Futch was also short with light brown hair. He always wore a smile that could break out into a belly laugh within seconds. Originally a Signal Officer and now a fixed-wing aviator by branch, he was also Special Forces qualified. I used to rag him, saying that it wasn't getting his flight pay that I resented, as he probably deserved that; it was his base pay that I had a problem with. Futch liked to run, and we hit it off right away. Everyone got along well in the group. Montero, Futch, and I buddied around and became great friends.

Futch and I were always planning our runs or talking about races we would do, and Monty got psyched about it. Early on he'd ask me about Ranger School and said he thought he'd like to attend. Right. Monty started running a little, but couldn't hang with Futch and me. Soon we noticed that he was losing a little of the pudginess. What was with that? Well, to make a long story short, Montero smoked Futch and me like cheap cigars on a 15K run at the end of the course. Where did that come from? Monty would go on to become the Transportation Officer for the Joint Special Operations Command at Fort Bragg and while there would attend Ranger school at Fort Benning, graduating at just over forty years old. He would also continue to rise in the Army, retiring as a Major General. This taught me to never underestimate a determined little Portuguese soldier.

As luck would have it, my old buddy Vaden Bessent showed up in the same class. He and Rose Marie lived one door down from us. Polly had gotten a job teaching. Often our classes would end early, and Vaden and I would sit around my house watching TV and swilling PBRs. Polly would come in and run us out. We would move to Vaden's. I'd stay there (normally a very short period of time) until Rose Marie ran me out. Ah, yes. The good times, they were a rolling.

Leavenworth offered a lot for the kids. Pollyanna and Tee had been on the swim team at Fort Polk. Both were natural athletes and excelled in every sport they tried. This was a two-edged sword: since they did so well with so little effort, both would get bored and never get overly excited about any one sport. Not enough to take it on to college anyway.

Both Tee and Pollyanna swam on the team at Leavenworth. I can still see them now pulling on their moon boots and tromping though knee deep snow to their waiting ride which would take them to swim practice at 0600 in the morning.

Tom and Nancy Futch had two boys, Greg and Will. Both boys also swam on the team. Weekends we would all pile into the Easter Egg and head out to the various swim meets around the state. The smell of chlorine and the punching of stopwatches consumed many a Saturday for us. We became close friends with the Futchs and remain so today.

NINE MONTHS LATER as graduation approached, I got the word that I was to be assigned to the 5th Special Forces Group as the Group S1. I was absolutely thrilled to be going back to Bragg and especially looking forward to another tour in Special Forces even if it meant I'd initially be the Group S1.

As it turned out, my orders sent me to the Personnel Management School at Fort Benjamin Harrison, Indiana, en route to Bragg. Much to my chagrin, this meant that I'd have to leave a little early and wouldn't be able to help Polly pack up our stuff in Leavenworth, clear quarters, and set up in Fayetteville, North Carolina, where we had bought a four-bedroom house in a very nice neighborhood. Fortunately for me, I had the greatest mother-in-law any son-in-law could ever, ever have. Ethelena would fly up to Leavenworth, help Polly pack and clear quarters, then ride with her and the kids in the Easter Egg down to Fayetteville. She'd help unpack and set up there. I know you may think your mother-in-law is great, too, but Ethelena was truly a saint. I loved her always and miss her so very much now since she left us at age 96.

LITTLE DID WE know when we arrived at Fort Bragg in 1982 that Polly and the kids would be at that one place for the next twenty-nine years. Both Pollyanna and Tee would go from the second and third grade to graduating from college while we were there. Polly would have a career in the Community College system and even earn her Doctorate from North Carolina State University. While for the most part I would remain assigned to Bragg, I would see service with Special Forces in Zaire, Turkey, Bosnia, Iraq, and various other spots around the world. Stand in the door!

Chapter 8

Fort Bragg, North Carolina-Again

S1/Deputy Commander Support 5th Special Forces Group

POLLY HAD ARRIVED well before me in Fayetteville, North Carolina, with her mother and the kids. By the time I finished up at Fort Ben Harrison and drove to Fayetteville, she had the house all set up and running fine. Wish I could have been there to help, but. . .. I drove out to Bragg and signed into post, then on to the 5th Group Headquarters. I reported to my new boss, the Group XO.

MAJ Bill Tangney, sitting behind his desk, waved me in when I knocked on the door. Bill was my height with rust colored hair, a ruddy complexion, and a quick smile. He was originally from New England but had graduated from the Citadel in South Carolina and adopted the South as his home. He still had a hint of the New England accent, which gave him away. Bill was married to Kathi, and a better command team there never was. Bill and I would serve together off and on for the rest of my career. He would also save me from myself more often than I care to count.

The Group Commander at this time was COL Time Bienheim. Bienheim would soon retire and was getting a head start on retired life. He mostly stayed in his office and painted watercolors, some quite good, and let Tangney run the Group. Things went well.

I had been the S1 for only about one month when one night my stomach began to hurt. I felt constipated, a condition that I rarely experienced, and took a laxative. Big mistake. I didn't get any sleep that night as things got worse. In the morning, I drove Polly to work in the Easter Egg and then drove straight to our dispensary. Fortunately for me, Doc Bill Fox, the Group Surgeon, was pulling dispensary duty. He took one look at me. Did a quick poking around my stomach then called an ambulance. I had appendicitis. By the time I'd gotten to the hospital and into surgery, my appendix had ruptured.

This wasn't the way I wanted to start out in a new job. I was laid up in the hospital for several days then sent home on convalescent leave.

I couldn't stand sitting around, so I went in to work. Tangney wasn't pleased that I'd done that, but what could he do. In my mind, I was sure if I didn't get back into my chair and soon, he would fill it with someone else. Couldn't have that.

BIENHEIM RETIRED AFTER I'd been in the Group for only two months, and in to replace him walked COL Jim Guest, my old Battalion Commander from Germany. A small world SF is. I always enjoyed working with Guest even though sometimes it was like riding a bucking Brahma bull wearing only your jockey shorts. He was a fountain of ideas, most all of which were way outside of the box. He would say, "Thomas," he always called me that, "What we need to do is. . .." and the fun began.

THE FIRST THING that Guest did was call Tangney and me in and say to me while looking at Tangney, "Thomas, what we need to do is organize in garrison as we do when we deploy."

When a SF Group deploys in support of combat operations, it does not use the S's to designate its staff. It is broken down into three centers: the Operations Center (consisting of the S3 and the S2), the Support Center (consisting of the S1, S4, and all special staff except the Signal Officer), and the Signal Center (consisting of the Group's Signal Officer and the Signal Platoon). Each center has a Director. This person, in the case of the Signal Center, is the Group Signal Officer. In the case of the Operations Center, it is the S3. In the case of the Support Center, it can be either the S1 or the S4. The Group XO is the rater or man in charge of the three directors. In our case, Guest declared that the S1 (read that me) would head the Support Center as its Director.

Actually, this was quite an insightful move. Brilliant even. Of course this would cause major personality conflicts when the S2, S4, and the special staff were told that they would no longer be directly under and rated by the XO.

Doc Fox howled the loudest when he found out that he would be rated by me, a lowly Major. I understood exactly how he felt. After all, I took on the official title of Deputy Commander for Support rather than Support Center Director as it would look better on my record. I was beginning to think about "my record" by this time

I proposed the change to Tangney, and he nodded in agreement and took charge of Fox. I liked Fox, and he had had a part in saving my life when I had appendicitis. Fox must have been on to something since he eventually retired as a Major General.

Another example of Guest and his "space shots" came when he took up the cause of creating a separate branch of the service for Special Operations. If approved by Congress, this would mean that Special Operations would be on the same level as the Army, Navy, Air Force, Marine, and Coast Guard. He argued that the individual services' Special Operations forces, i.e. SEAL, Air Force Special Ops units, and Army's Special Forces, Civil Affairs, and Psychological Operations had unique requirements and were being discriminated against within their separate services. He was right, of course.

This suggestion, however, brought wails like you never heard from the services, mostly Army and Navy, with the Army being the loudest. The battle was on, and position papers were flying around the Army. Soon, some in Congress took up the cause of this "sixth" service. That's when the Army got off the dime and made its move. To make a long story short, the Army decided to make Special Forces a combat arms branch equal to the Infantry, Armor, and Artillery. In my heart I thought that was what Guest was aiming at all along. Who knows?

One day I brought a paper in to Tangney with his signature block on it. It was an official request for Astronaut Wings. It was routed through 1st SOCOM (Special Operations Command) to the Awards Section in MILPERSEN. I placed it on Tangney's desk. "I'm putting us in for an award."

Tangney sat back in his chair, picked up the paper, and began to read. A smile broke on his face. "Okay. I'll bite." He motioned "Give me" with his hands.

"Well, we've been on so many space shots with the Old Man, I figured we deserve Astronaut Wings."

Shaking his head, Tangney smiled, wadded the request up, and slam dunked it into his grey metal trash can.

Noting ventured. Nothing gained.

I WAS KNOWN for speaking my mind. Some might say I'd put my mouth in the run position before my brain was in gear. Whatever. Anyway, the XVIII Airborne Corps commander, LTG James Lindsay, who had at one time served in Special Forces, was being briefed by MG Lutz, the commander of 1st SOCOM (Special Operations Command) under which fell all the SF Groups, active and National Guard. Lutz was a great leader. He was a Calvary officer, but had an exceptional background in Special Forces. He had commanded both A and B Teams. He was now the commander of 1st SOCOM. Attending this meeting with Lindsay were all of SOCOM's primary staff including the 5th and 7th Group's S1 (in my case, I was the S1 and Deputy Commander for Support), and 5th and 7th Group's XOs.

Lindsay had a program called the 2+2. A normal tour at an Army post for a captain at the time was three years. SF hadn't formed a branch yet, and all the SF Captains who were branched Infantry had to have company command time in order to make Major. Lindsay knew this, and in order to help the SF captains out, he arranged for the Department of the Army to make an exception to the three-year rule. This exception allowed a captain to spend two years in the 82nd Airborne and two years after graduating from the SF Qualification Course with Special Forces. SF captains, in turn, would spend two years in SF after graduating the Q Course, then be sent to the 82nd to get their one or two years of company command time.

Well, you'd think this would be a good deal. Right? It was, and it wasn't. It was a good deal since we could select our best captains who had performed exceptionally well commanding Teams and reward them with a chance to "get their ticket punched" with Infantry branch by getting command of a company in the 82nd. And this is what we did.

However, when it came to the 82nd sending us their most successful company commanders to serve in SF and get Team command time, the system broke down. Normally, what we got from the 82nd were captains who had not performed so well as a company commander. I believed that the 82nd was using the program to dump captains who had stumbled, and the 82nd wanted to move them out. Of course, this wasn't the case with all captains that the 82nd sent us, but it was my observation as the Group S1 that this occurred more often than not.

MG Lutz held the meeting in the large conference room. Even so, there was standing room only. Being a lowly Group S1, I was relegated to standing near the back. Tangney sat ass to ass with MG Lutz's staff and the 7th Group's XO around the long table.

At the end of our briefing, Lindsay leaned back in his chair and said, "Well, guys, tell me how the 2+2 program is doing." He followed that with a big smile, knowing he'd hear what he wanted to hear. Not so fast, Mojumbo.

Before anyone else had a chance to answer, my hand flew up. Tangney, who I'd often complained to about what the 82nd was doing to us with the 2+2, tilted his head back, lifting his eyes to the ceiling. He knew what was coming.

Lutz, who I'd met on several occasions, nodded at me and said, "MAJ Davis, what are your thoughts?"

Tangney now dropped his head, chin to chest as though he was praying, and I guess he was. It wouldn't do him any good, by the way.

"Well sir, It's like this. We send the 82nd the very best captains we have. Those who have been Team Leaders of some of our best teams. What we get from the 82nd in return are their duds."

It was the God's truth, and everyone in the room on the SF side knew it. A chilling quiet seared the room.

Lutz tried his best to look taken aback with the very idea that one of his majors would stand in front of the three-star general commanding the corps and say something so blunt. He might have pulled it off except his mouth kept twitching from his fake frown into a quick but genuine smile. Tangney began slowly shaking his head. Others sat either stunned or doing their best not to throw me a high five.

Lindsay, clearly taken aback, gave a little cough then turned to his aid who sat opened mouthed and said, "Make a note of that. I want my G1 to take a look at all the records of the captains that the 82nd has sent over here."

The aid scratched furiously in his little green notebook.

Well, that pretty much wound things up in the briefing. All were more than happy to stand when Lindsay, followed close behind by his aid, walked out of the room.

All Lutz did was shake his head at me and smile. I smiled and shrugged, flipping my hands up in a "What else could I do?" gesture.

On our way walking back to the Group headquarters, Tangney said, "You just had to do it. Didn't you?"

"Well, the guy asked the question. Didn't he?"

Tangney just smiled and shook his head.

When we got back to the Group, Guest had already gotten the word. He was always shaking the trees, so to speak, when it came to challenging the system, and that was just what had I done. He loved it. I knew he would. He had spent several tours off and on with the 82nd. I had heard him often refer to the 82nd as "The Imperial (emphasis on imperial) 82nd . . . AIRBORNE."

The quality of captains coming to us from the 82nd increased greatly after that meeting. Must have been a coincidence.

XO 1/5 SFGA

THE FIRST BATTALION of the 5th Special Forces Group had just changed command, and a guy named LTC Dan Edwards stepped into the position. I had been the Group's S1/Deputy Commander for Support for nine months and was looking to change jobs. I needed to be a battalion XO in order to punch my ticket for Lieutenant Colonel and hopefully command a battalion myself. The XO position in 1/5 was open. I asked Tangney if he would support my leaving to be the XO there. He said he would, and we took it to Guest, who readily agreed. He knew how important the job would be for me.

My next stop was to see Edwards. He was a North Carolinian from the mountains. Rifle barrel straight, he held secure to his principles. He stood a few inches shorter than me with a wide face and wide shoulders. He ran some and weight lifted some to stay in fighting shape. He played the guitar well, but was shy about demonstrating his ability. In fact, I don't believe I ever heard him play more than a cord or two in the many years that we have been friends. He married Carolyn, who also grew up in the western North Carolina mountains. Polly and I became very close to both. As always in SF, I would work for Dan again. The next time in Korea.

Dan had someone else in mind for his XO but was aware that Guest wanted me in the job. He relented, and a week later I reported in to the battalion. SF's becoming its own branch of the Army was still four years away. But we were one of the Army's Special Operations units. Special Operations is an umbrella term that includes Special Forces, Rangers, Civil Affairs, Psychological Operations, Special Operations Aviation, and a Support and Signal Battalion, all of which fell under the 1st Special Operations Command. The news media did then and still often does today refer to Special Operation Forces as Special Forces which is factually not correct.

It was early in 1983, and Army money was not flowing our way. I remember that the Group gave our Battalion only $20,000 in training funds for that entire year. Edwards wanted to run a battalion-level FTX in the southwestern desert. This would take most of the 20K, leaving only $500.00 to each of our three companies to spend on discretionary training. In today's Special Forces, a 12-man A Team would run through $20,000.00 in a month or two.

CPT Mark Phalen, an exceptional officer whose attention to detail was superb, served as my S4. Ever since I commanded the mechanized

Infantry company, I took a special interest in supply accountability. This was something lacking in most SF units, and 1/5 was no exception. After a little coaching and counseling, Mark had every piece of equipment signed for down to Team level. He did so well with that, I put him in charge of all monies the battalion had at its disposal. Mark would progress up the ranks eventually retiring in the General Officer ranks. No surprise there.

As XO, I was responsible for running the staff while the company commanders fell under Edwards. After a couple of months, I thought things were running along pretty smooth. I was death on suspenses, and kept a list on that yellow legal-size paper on my desk, listing the staff section responsible for a suspense, a short description of the suspense, the date the suspense was due, and, finally, the date we met the suspense. During our weekly staff meetings I would run down the list and get updates from each staff officer. All the staff officers except the Major, who was the S3, slipped into my office and checked the list to make sure they knew what and when they had something due. The only staff officer who missed suspenses on a regular basis was the S3. Fortunately for me the assistant S3 was an extremely competent lieutenant named Mark Payne. Payne would soon become the Battalion's S3.

Edwards had the ability to put things into perspective in the most profound way. I remember one time Lt Payne came into my office with a tasking that had come through the Group that was absolutely unreasonable. I told Payne to follow me, and we walked into Edwards' office. He had his half-frame black horn-rimmed reading glasses perched on the tip of his nose. He always wore them when he read anything. He looked up at me then at Payne. "This can't be something that's going to make me happy." He gave us his usual smile.

I placed the tasking paper down in front of him and stood back. "You're not going to believe what we just got passed down from Group. Looks like it originated at 1st SOCOM, so you know it is full of it." I stepped back and looked out the window at the cars passing by on Ardennes Street. Edwards perused the tasking.

"We should fight this," I said.

"Well, we can, but we'd lose." Edwards slowly shook his head while pushing out his lower lip.

"They want us to send a whole company out in support of Robin Stage." Robin Stage was an exercise that the Special Warfare School conducted as part of SF training. "And they want them out there tomorrow! That's crazy." We were in detail support for the next thirty days which meant that we got all taskings from any unit above us. "Most of our guys are out doing something else. Don't you think they could have given us a few more days notice? It's impossible." I crossed my arms in front of my chest and frowned.

Edwards just smile; then he said something that would resonate with me and guide my actions as I climbed up through the ranks to Colonel. "Nothing's impossible for the one who doesn't have to do it."

I slowly nodded as what he said sank in. Payne and I turned and walked out of the office. We knew we had to face one of the company commanders who would tell us what we already knew–that we were crazy. But what else was new?

SOMETIMES A GOOD kick in the ass is just what the doctor ordered. I had a bad habit of cutting in on Edwards while he was talking. I wasn't intentionally being rude. I would just have a thought that was pertinent to the conversation and would interject it without thinking. I did that with a lot of people. Edwards would hold his hand up and say, "Would you just let me finish my thought before you jumped into the conversation?"

I did this one too many times.

During a conversation I interrupted him, and he held up his hand. "Take a break. I need a minute." Edwards waved me out of the room.

I shrugged and left.

In about fifteen minutes, he called me back in and motioned me to take a seat. I did.

He had his reading glasses on and a piece of paper in his hand. It was a written counseling statement. He read it to me. Basically, he told me in writing what he had been telling me verbally for some time. I listened to him read without interrupting for once. When he finished he looked up and said, "You got any questions?"

I shook my head "no," and he told me to sign the form and gave me a copy.

Then he smiled and said, "That'll be all."

I walked out of his office a wiser and more humble guy. It was the best thing he could have done for me. From that day on I never interrupted him again, and for the most part never interrupted anybody else. Well, almost never.

EDWARDS WAS A man of great principle. One time one of our Team Leaders who was on a MTT (Mobil Training Team) mission to Africa did something really dumb. Contrary to standing orders and good common sense, he had drunk purified water but put local ice in it. That did it. He got so sick he had to be evacuated. Guest was pissed to say the least, and wanted Edwards to relieve the guy.

I argued that the Captain was a dud, and this would be a good excuse to get rid of him. Relief for Cause is serious, especially if the officer getting relieved is in a command position. I thought it was long over due for this guy.

Edwards wasn't so sure. "Tom, I agree with you he isn't the shiniest ornament on the tree, but I just don't believe that I should relieve him for this."

"Guest won't like it." The one thing about COL Guest was that, of all the people around, he was the very last guy you wanted to piss off. Even I took great care not to do that. And that was really saying something.

Edwards wouldn't budge on it, and told Guest so. This impressed me to no end. I would always remember what he did, and one day as a Battalion Commander I would do something similar for one of my guys.

MY YEAR AS XO under Edwards had drawn to a close. I had learned so very much from him, and Polly and I had found friends for life in him and Carolyn.

Tangney had moved out of the Group XO job and up to 1st SOCOM to be the SGS (Secretary to the General's Staff).

My alternate specialty was 41 or Personnel Management, so when a slot as Deputy for the 1st SOCOM G1 opened up, Tangney recommended to the Chief of Staff that I be selected for the job. I'd had my mandatory year as an XO, and it was time to move on. I would spend a little over three years in as the Deputy G1 and make many great and wonderful friends and one very powerful enemy.

Deputy G1, 1st SOCOM

AS THE DEPUTY G1 I worked directly for COL Ken Rice. Rice was small in stature and large in heart. His voice had the slight rasp of a smoker. His long sad face reflected the riggers of a recent and bad divorce. The sadness would vanish when he met Elta within the next year. Rice had commanded a battalion in the 82nd and then became the

Division's G1. As a result, he was well versed in the burdens of command and the frustrations of the staff. I learned a lot from the man and enjoyed working for him.

Since Rice was not Special Forces, he often sought my advice when it came to matters that would affect the four Special Forces Groups. For the most part, I believe I gave him sound advice or, if not that, at least an honest opinion. For the most part, he acted on that advice.

Rice's direct boss and the next above me was the Chief of Staff, COL Sydney Shacknow. As time would pass, I would interact more and more with Shacknow. Shacknow was a few inches shorter than my six foot one, had dark hair cut close on the sides and combed to the right on top. A man of tremendous foresight, he had the kindest eyes I've ever seen. This was probably because those eyes had seen the very worst and the very best in humanity. He spent his early years in an Nazi concentration camp, eventually making his way to the States. He worked his way from Private to Sergeant First Class and from there into the officer ranks where he would retire as a two-star general. His memoir, *Hope and Honor*, is a must read. Sid would also be the only man to ever see me cry.

I would have friends in high places in SOCOM headquarters. LTC Bill Tangney was Secretary to the General's Staff. My old boss from Bad Tolz, COL Don Solan, was the Deputy Commander. Both would come to my rescue more often than they probably should have.

SHORTLY AFTER ARRIVING in the G1 shop, I had the opportunity to attend a one week class, Personnel Management for Executives, in Atlanta, Georgia. It was the best single week I'd spent in a classroom since signing up to serve Uncle Sam. I walked away firm in a theory of management that also would apply to command.

The Five Principles of Good Management would influence the rest of my career in the military. I adopted them as the Five Pillars of my Command Philosophy. I will, because I must, give you the *Readers Digest* version of them now. You will see them again many times from this point to the end of this fantastic read.

THE ARMY HAS long debated whether a good commander must also be a good manager. Do you have to be both to be either? If so, are they of equal importance? And so on. For the sake of argument, let's say that good leaders must be both. In light of this, five principles of good management emerge from the shadows and beg an introduction.

1. CONSISTENT. A good manager is consistent. Webster's Dictionary states that consistency is "always acting or behaving in the same way; or, harmony of conduct or practice with profession." To us, this means, among other things, that once you chart your unit's course, you don't make 90 and 180 degree turns every other month. (Tweaks of a few degrees are not only acceptable, they're necessary.) Another example–What's deemed as unacceptable behavior today will be deemed as unacceptable behavior a month, a quarter, or a year from now. Basically, you react to similar situations in the same way.

2. IMPARTIAL. A good manager is impartial. Again, let's refer to Mr. Webster. He defines impartial as not partial or biased; or, treating or affecting all equally. This principle is easily understood, but consistently violated. Let's say your policy is that everyone in the unit will attend PT (physical training) Monday through Friday at 0600. That's too early? Make it 0630. The policy further states that those not attending morning PT, for whatever reason, will report to duty by 0730. You enforce this on everyone except SSG Shootumup, your AMMO (ammunition) sergeant. He doesn't participate in PT at 0630 and doesn't come to work until 0830. You're being partial to him. You're violating this principle. The bottom line is that you must insure that no one in your unit receives preferential treatment.

3. WARNING. A good manager gives warning–first oral, then written–prior to taking action against someone. We Army types are doing good with the first part–oral, but we're in the dumper on the second part–written. SSG Slipinthemud has missed several formations over the past six months. His company commander gave him an Article 15 (non-judicial punishment given by a company or field grade officer) for his most recent absence. He has appealed it to you, the battalion commander. Since the Army has established mandatory written initial and quarterly counseling for noncommissioned officers, you ask the chain of command to provide you with Slipinthemud's file. Guess what–no mention of him ever missing a formation. Wrong.

4. TIMELY. A good manager is timely when taking action for or against someone. A letter of reprimand or commendation delivered to CPT DoWrong/DoRight three months after the act loses its effect. While it's true that the statute of limitations on an Article 15 is two years, waiting to give one to a soldier six months after the fact is hardly the right thing to do. Remember the old saying: Justice delayed is justice denied.

5. DO versus SAY. What a good manager says must be very, very close to what he does. If you say you will enforce ALL Army regulations and you don't, or if your field SOP (standard operating procedure) states you will do such-and-such a certain way, but you don't, you're violating this principle. Get the picture?

And there you have the five principles of good management standing before you in the spotlight. Exception to any one or combination of these principles is strictly forbidden. Do you buy that? You do? Well, guess what? You're wrong.

Not only is it acceptable to make exceptions to these five principles, but at times it's downright necessary. However, there are a few things you must do before an exception is made. First, make sure the situation really warrants this exception. Next, let your soldiers know you're making an exception. Finally, explain to them why you're making it. You aren't asking for their permission. You're not asking for their approval. You're simply informing them of a decision you've made and why you've made it.

An example: Remember SSG Shootumup, the AMMO sergeant you're not requiring to make the 0630 PT formations and you've allowed to come in late to work? He's just gotten divorced and gained custody of his 9-year-old son. He needs the morning PT time to get his son off to school. You've decided to let him do PT during the lunch hour, so that's what he's doing. Your response could go something like this: "I'm making an exception for SSG Shootumup. He's not required to participate in morning PT, and he doesn't have to come at the normal time. I'm doing this because of his unique family situation. He'll do his PT during the lunch hour." This doesn't have to be announced in a unit formation. You could put it out through NCO channels. Word will get around, and those who had the perception that the "Old Man" makes everyone except his AMMO NCO do PT in the morning will understand. (Of course, if you do this for SSG Shootumup, you must be prepared to do it for anyone else in a similar situation. Remember, consistency.)

Throughout the rest of this memoir, you will see how these five principles guide my actions in both staff and command positions. For the most part, they would keep me out of trouble, but, occasionally, the reverse would happen. It's tough adhering to these principles when someone above you doesn't.

The Most Fun I Ever Had With My Clothes On — 233

WHILE SERVING AS the Battalion XO, I kept up a good physical fitness schedule. Time permitting. Now that I was in the G1 shop and had more flexible hours, I went at it with a vengeance. I would come onto post early and swim a mile in the pool at Lee Field house. At lunch I would bike twenty miles, then run two miles, preparing me for the Triathlon transition from bike to run. I varied the workouts, sometimes longer and sometimes shorter. On the weekends, I'd take one day for a fifty-mile bike or a eight- to ten-mile run. The other weekend day, I'd rest.

Bob Glanville showed up at Bragg and signed in as the XO for the 7th SF Group. He and Diane bought a house two doors down from ours in the Van Story subdivision. We entered the North Carolina Triathlon circuit and spent many a happy weekend pounding the pavement.

I SAT AT my desk in my office located across the room from Rice's, looking out the window and across the parking lot. A pile of paperwork begged to be addressed. The phone rang, and I picked it up. "Major Davis."

"Tom. Frank. Need a favor." Frank Tony, a major and company commander in the 5th Group, and I were friends. There were two schools of thought on Tony: either you liked him or you didn't like him. No one was indifferent. Whenever the subject of Frank Tony came up, I was quick to defend him, saying to everyone present that the one thing about Tony that I really respected was you never had to worry about him stabbing you in the back. Now, the chest was another matter all together. Another thing I liked abut Tony was that he never beat around the bush. He just puled it up by its roots. He expected no slack and gave none. He stood eye to eye with me. Having spent quite a bit of time in the Ranger Regiment, he kept his hair cut high-and-tight sporting bare whitewalls all around and a flat top so short that a bald spot showed on the very top of his head.

"Okay, Frank. Whattaya need?" I turned around to my desk and readied a pen on the yellow legal pad I kept there.

"I got this young captain working for me. He's just gotten a divorce and has full custody of his four-year-old. Name's Jacklyn. Cute little thing. Anyway, I want you to find him a job where he doesn't have to go TDY (Temporary Duty–short one to three weeks away from the headquarters). He needs a stable job that'll give him time to set up something to deal with being a single parent."

"Who is this and what have you done with Frank Tony?" I smiled at the thought of Frank showing a soft side.

"No, really. The guy's got unlimited potential. He loves the Army and especially SF. He's going far. Problem is that he's getting out if we can't work a deal for him. Now, don't tell him I'm asking you to do this. He wouldn't want special treatment. He's serious about getting out. I think he's filling out the paperwork now."

The fact that Tony was standing up for this guy was impressive, but even more impressive was the selflessness of what the guy was about to do to take care of his little girl. "Gimme his name."

"Gary, but he goes by Mike, Jones."

I wrote it down. "Tell him to hold off on that paperwork until I can look into this."

The G1 shop was broken down into separate sections. The Officer Management section was headed up by Majors Joe Whitley and Rick Ballard. Ballard would one day save my bacon, a story for later. Also in that section was Chief Warrant Officer, Russ Voner. Voner managed all warrant officer assignments in SOCOM. In the plans shop sat Ron O'brian, who also was in charge of monitoring/staffing all SOCOM regulations and enforcement of Army regulations. Hanging out there with no one but me to supervise them were the Safety Officer, Equal Opportunity NCO, and a few other special function personnel. What I needed was a guy to take the management of regulations from O'brian and all the other folks from me. I'd call him the Proponency Officer. And this is what I pitched to Rice. He agreed. Next I called in Whitley and Ballard and laid it out for them. They were on board.

I called Tony back and asked him to have Jones come see me. I told him what I had in mind. As it would turn out, Tony was dead on about Jones. Both he and Tony would retire as General Officers.

"Sounds good. But what about TDY and PT formations. Crap like that?" Tony said.

"Well, we don't have PT formations around here. Everyone is on his own for that. So far as TDY goes, if there's a need, either I'll go or have someone else go in his place."

"Sounds good, but don't tell Jones that."

I agreed.

THE NEXT DAY a knock shook my door. "Come," I said. And in walked a guy who stood about two inches above me with shoulders three times as wide as mine. He had a slightly receding hairline that would reveal more as time went on. He'd played football at LSU and still looked like it.

"Sir, Captain Jones. Major Tony told me to stop by and see you about a job."

I'd made some good decisions in my life and some bad ones. Hiring Mike Jones to work for me was the third best decision of my life. The first: marrying Polly. The second: accepting Brockelman as my Team Sergeant on the SCUBA Team. And this one.

I explained what I wanted him to do and asked if he was willing to give it a shot. He gave me a wide smile and big "HOOAH." The deal was sealed.

Not only would Jones be an extremely productive member of the G1 team, but I'd drag him into triathlon training with me. We would share some exciting adventures. In time he'd meet a great lady. He and Helen would marry. And I'm proud to report that they're still happily married today.

A NEW CG took over command of 1st SOCOM, MG Leroy Suddath. He was not SF qualified, but had extensive experience as an infantry officer in both conventional units as well as the Rangers. His head was capped with white hair even though he wasn't all that old. He stood a good three or four inches above me and was long in every respect–legs, waist, neck, face, arms, and fingers. He had to have had the longest fingers I've ever seen on a person. Think ET. He was from Georgia, specifically, Savanna. His voice was deep, rich, and sincere. I was leery of him for quite a while but eventually came to admire and respect him.

Suddath's first decree was, "We will enforce ALL Army regulations." It wasn't that we in Special Forces disregarded Army regulations as a matter of habit. Through my many years in SF, I'd seen some serious cutting corners in that respect. Given the Special Forces' missions, sometimes Army regulations presented additional challenges in accomplishing them.

The G1 is the watchdog when it comes to enforcement of regulations. Okay. In my mind, ALL meant every single one. No exceptions. There was just one little problem. MG Suddath, who was not SF qualified, walked around wearing a Green Beret..

I called Jones into my office. "Jonesy, go check the reg and see what it says about what type of head gear we here in SOCOM are supposed to be wearing."

Jones smiled and nodded that he would. He knew where I was headed.

Within the hour he entered my office with Army Regulation 670-1–Wear And Appearance Of Army Uniforms And Insignia and the pertinent SOCOM regulation in hand. "Okay, boss. Here's how I read it. To wear the Green Beret in the headquarters, you have to be SF qualified and be in a slot designated Special Forces."

"And so far as the CG is concerned. . .." I lifted my eye brows in anticipation of his answer.

"Well, the CG *is* in an SF slot, but he *isn't* SF qualified. He should be wearing a maroon beret since he is airborne, and we're an airborne unit."

I sat back in my chair and steepled my fingers. "So if we're going to enforce ALL Army regulations. . ."

"The CG is out of uniform and in violation of 670-1." Jones finished my thought.

My next stop was Rice. He didn't want to touch it. Absolutely not. No way. Period. That was it. End.

"Well, we're not being consistent. Also what we're saying we'll do is not close to what we are actually doing," I reasoned.

"Not that again. I should've never sent you to that personnel management thing in Atlanta." Rice shook his head and sighed.

He wouldn't budge. I asked permission to take it to Shacknow, the chief of staff. To his credit, Rice waved me on.

I stopped by Tangney's office. He was Secretary to the General's Staff. I needed to see him before I could take the issue to Shacknow. While I was explaining to Tangney what I wanted to do, Solan walked in. Both thought I'd lost my mind, but gave the go ahead.

Shacknow was always very indulgent with me. He listened patiently to what I had to say. At the end, he said, "Is this really a battle you want to fight?"

I thought about it. "He did say we would enforce ALL Army regs. This is an Army reg. So yes."

"Okay. I'll discuss it with the CG." Shacknow wasn't very happy. "You know, Tom, you must sleep well every night seeing how everything is so black and white with you."

This gave me pause. Could there be some grey here? Maybe. Maybe not. I decided not.

I don't know how the conversation went between Suddath and Shacknow. I do know that Suddath kept wearing a Green Beret.

IT WASN'T LONG before the Army came out with a no-smoking regulation. It declared that soldiers could only smoke in designated areas. You guessed it. The CG smoked, and he smoked in his office. At my request, Jones brought me the reg and with it in hand I retraced the steps I'd taken with the beret issue. Rice smoked but he always did so in one of the outside smoking areas. Same with Tangney.

Soon I found myself face to face with Shacknow. He shook his head. "You just don't give up do you, MAJ Davis." It was "Major" on this one, not "Tom."

"Sir, it's right here in the reg." I pointed my finger at the regulation I had placed on Shacknow's desk.

"Well," countered Shacknow. "The GG has designated his office as a smoking area."

I shook my head. "No can do. The reg specifically states that you can't designate your office as a smoking area. And if the CG wants to play it that way, then we should let everyone in the command know that they can drop in and puff away there. Think that'll fly?"

Shacknow smiled and shook his head. This time he knew I was right. No question about it. Well to shorten the story, soon I noticed the CG enjoying his smokes in one of the designated areas. And pigs do sometimes fly.

BG WAYNE DOWNING became the Deputy Commanding General of 1st SOCOM. Downing was a Ranger through and through. He had commanded every level in the Rangers from Platoon through the Regiment. While doing nothing overt, he was not a friend of Special Forces. I didn't like him.

COL Rice and I sat in Rice's office waiting to give Downing the G1 briefing. Downing strutted in, black beret in hand, and sat down. He looked at me then at Rice. "Okay, Ken. Whatta ya got for me?"

Rice flipped through the printed PowerPoint slides we had prepared. I sat wondering why I was even there. Rice finished and asked for questions. Downing had a few, then concluded, saying, "Ken, Major General Suddath has said that we are going to enforce all Army regulations. I want to make sure we're doing that. To the letter. You guys are the gatekeeper for this." Rice nodded and Downing continued, "So how're we doing?"

I couldn't stand it. Downing had been wearing his Ranger black beret since he came to the headquarters. He was in the same position that Suddath was. The DCG slot was designated for Special Forces. If the guy filling it wasn't SF qualified, he would have to wear the maroon beret of the Airborne. The only slots coded for Ranger in SOCOM Headquarters were the Ranger NCO and Officer liaison in the G3. Those guys were authorized to wear black berets.

Rice squirmed in his chair. He knew Downing was in violation of the uniform reg when he wore the black beret, but didn't want to make waves. I had no such qualm.

"Sir. If I may." I looked at Rice, but before Rice could tell me to shut up, I went on. "I'm glad to hear you say that. Actually, you are in violation of the uniform regulation. Only individuals assigned to a Ranger slot within SOCOM are authorized to wear a black beret. Your slot isn't coded for Ranger. You can, however, wear a maroon beret." I broke into a big smile.

Downing was taken aback. He turned and flashed a frown in my direction. "Well...uh...Major Davis is it?" I nodded. "General Suddath has authorized me to wear this beret. I'm a Ranger, and Rangers are an important part of this organization."

"Sir, with all due respect, General Suddath doesn't have the authority to authorize anyone to wear a uniform that is not approved in the reg." Jones had researched this for me. "The only individuals in the Army that can designate their own uniforms are the Army Chief of Staff or a five-star General of the Armies."

Rice looked like he wanted to be anywhere on earth except sitting where he was. Downing mumbled something about we'd see about that; then he stood and stomped out of the room.

Rice, shaking his head said, "Tommy, you just couldn't keep your mouth shut for just two more minutes could you?"

I shrugged my shoulders and turned my hands up in a "What else could I do?" motion.

Rice smiled and waved me out of the room.

It was the first time I poked Downing in the eye, but it wouldn't be the last. Not by a long shot.

I WAS WELL into my first year as the Deputy G1 when I dragged my brother, John, with me down to Fort Lauderdale, Florida, to watch me compete in the Penrod's Triathlon. We stayed with my aunt and uncle, Ben and Genie Ash. They were the parents of Benny Ash, who had somehow found me out in the middle of nowhere in Denmark. My team was attending the Danish Combat Swimmer School at the time.

After the race, John got so excited about the sport he declared that he wanted to participate in the event next year. I set him up with a training schedule and he had stuck with it. As of the writing of this memoir, John has logged a total of 21,081 miles running. Hooah!

THE NEXT YEAR I convinced Mike Jones that he was ready for prime time and that he should go down to Lauderdale with John, my son Tee, and me. To compete, you had to be at least fifteen. I had secured a fake ID for Tee, indicating that he was fifteen when he was actually fourteen. What can I say?

The swim starts the Triathlon. All swimmers begin at the same time, and it's chaos as the faster swimmers push to get out front, kicking and literally crawling over one another. If you stop or hesitate within the teaming mass of arms and legs, you drown or just about drown. Whatever you do, you must keep your forward momentum going, grabbing air when you can. I had prepared Tee for the swim by swimming laps in the Sports Center's pool. I swam alongside him in the lane, constantly swimming over his back and pushing him under water. We'd do this in sections of fifty yards as a time with a short rest between laps. As a result Tee was ready for what was to come. I had done the same with Pollyanna when she was preparing for the swim portion of Triathlons to come.

Penrod's Tinman Triathlon consisted of a one-mile ocean swim (in the surf), 20-mile bike, and 10K run. All events paralleled the beach. I thought it'd be a good idea if we went for a swim the day before. Jones and John hit the water and just about drowned. The surf was up and snarling, but it would be even rougher the next day of the event. Add the surf to the mass of arms and legs thrashing the water around you, and you had one hell of a challenge ahead.

A storm was brewing when we started the swim. Tee and I finished before the worst hit. John and Jones barely made it out before the rain and wind ripped across the coast, and race officials terminated the swim for anyone who was still in the water.

The bike ride was two five-mile down and five-mile back loops up and down the highway that paralleled the beach. When peddling into the wind, it was hard to hold a ten-mile-an-hour pace. When we made the turn around and had the wind to our backs, the pace picked up to forty-five miles an hour.

The rain came in blankets. We couldn't see the stop lights until we were almost under them. The last time I had seen rain close to that was the monsoons in Vietnam. And this made the monsoons look like an April shower.

The transition area where we changed from the swim to the bike, and bike to the run, flooded. Running and biking shoes, gym bags, socks, and anything left there floated out into the street. Of the many Triathlons I had done in the past, this was the most challenging. I'm proud to say that we all finished the race none the worse for wear. And, as always, good times were had by all.

TIME MARCHED ON as time will do. I sat in my office, plowing thought the hated and ever-growing pile of paperwork. The semiannual physical fitness test was coming up in a week. No big deal. No big deal until the Headquarters Company Commander, CPT Tiree, poked his head in my office. "Sir, can I have a word?"

I waved him in and motioned for him to sit. "What's up, Mr. Company Commander?"

I'd told Tiree when he took over the company that I'd been in his shoes, and that if he ever needed anything from me or my folks he would get it. Big mistake.

"Sir, I need a favor." This was never a good sign. "You know the PT test is coming up next week."

I nodded that I did.

"Well...It's like this...You know the last one we had... Well some of the guys were complaining about how it was administered." His eyes begged me not to push for details.

I knew well what was coming. During our last PT test, BG Downing, who had never not maxed a PT Test in his life, got credit for doing over 100 push-ups and 100 sit-ups in the two minutes allotted for each. The problem was that few, if any, were to standards. He barely nodded his head and received credit for a push-up. The sit-ups were done no better. The grader just kept counting, giving him credit. The rest of the company had to perform all to standards. Many of the lower ranking soldiers complained bitterly to the Company Commander.

"I get it. You want me to grade the Command Group." The Command Group consisted of Downing, Suddath, Shacknow, Tangney, and Command Sergeant Major Homestock.

"Well, you're the only one whom we all know that will grade everyone to standards." This was pure stroking. What he was really saying was that I was the only one who was stupid enough to hold Downing to standards.

I said I would, knowing what was coming. I have to write here that Downing was in really great physical condition, and I admired him for that. The problem was that he had never been graded to standards. If he had, he would have known what to expect. There is no question he could have maxed all the events without drawing a deep breath.

On the day of the PT test, Downing walked up and said, "Tom, what are you going to do, smoke us?"

I said, "No sir. Just going to give the test by the book."

It was the first PT test BG Downing took in his military career that he did not max. He passed but fell well short of max, in both the sit-ups and the push-ups. When he got up from doing the sit-ups, he jabbed his finger at me. "Next time we have a PT Test, I'm going to grade you."

Not being able to keep my mouth shut and remembering another Edwardsism (Never interfere with a soldier who's doing his job), I shot back, "Sir, let's not wait. Test me now." He stormed off.

In all fairness, I must state for the record that he did do the two-mile run in twelve minutes and a few seconds. An excellent time. When I said he was in great shape, I meant it.

I completed grading the rest of the command group. All passed. Of course, SGM Homestock was the only one who maxed everything.

All had a great time watching the spectacle, and it would turn out to be one of Tangney's favorite "Davis stories." I'm glad I could provide him with some comic relief. He would soon leave the SGS job to take command of the 3rd Battalion of the 5th Special Forces Group here at Bragg. He wouldn't be far away, and that was a good thing for me.

The time would come years later when Downing would extract his pound of flesh from me. Or so he would think.

IN APRIL OF 1986, LTG James J. Lindsay was still serving as commander of the 18th Airborne Corps and Fort Bragg. Lindsay was in fantastic physical condition. He ran every year in the 20-Mile Long Street run and regularly participated in Triathlons around the state. He was very much into physical fitness. As a result, he directed the Installation's DPCA to conduct tryouts for a team he wanted to represent Fort Bragg and the Army at the National Fitness Classic V. This year Houston, Texas, hosted the event. Numerous corporate headquarters sent eight-person teams to compete. We would represent the Army. This year a total of thirty-six teams would participate. A team consisted of two females and six males. I tried out for the team and was selected. Since I was the ranking soldier, I was dubbed the Team's

Leader. All members were in great shape, so they trained on their own for the events they would compete in.

We did well. On our return, I received a Letter of Commendation from LTG Lindsay. His letter read:

```
1. You are commended for your outstanding performance of
duty on 2 and 3 May 1986 as a participant in the
National Fitness Classic V in Houston, Texas. This
nationwide competition was sponsored by the President's
Council on Physical Fitness and Sports. You were
selected as one of eight individuals to represent Fort
Bragg and the United States Army.

2. As a direct result of your efforts, the team placed
1st out of 36 teams in the Classic TRIATHLON AND THE
200-Yard Swim Relay, 9th in the Stationary Cycle and 6th
in the overall standings.

3. The professionalism, dedication to duty, and positive
attitude you displayed have truly set you apart from
your peers. You are encouraged to continue to perform in
an exemplary manner.
```

I WAS INTO my third year as the Deputy G1 when my name popped out on the list to command a battalion. I was stunned and most excited. I was scheduled to take over 3rd Battalion of the 5th Group. Tangney currently commanded this unit. The only problem was that MG Suddath insisted that all officers selected to command an SF Battalion would attend language school to study the language pertinent to their targeted area. My area would be Africa south of the Sahara, and the language was French. This is one of the many languages I had tried and failed to pass throughout my life. I was looking as forward to this as I would a tooth extraction. Without Novocain.

IT WAS NOW late in 1986 and things were going well for Polly and me. Then came a blow that tore through my gut like a rusted KA-BAR. Polly's vision began to fade then almost went out. She experienced numbness in both legs, and the loss of hearing in one ear. After several visits to the doctor, we found out that she had Multiple Sclerosis, MS.

When I got to work the day after we received the news, I told Rice what was going on. Not surprisingly, I wasn't thinking clearly. He told me to sit down with Shacknow and talk. I was scheduled to take command of the battalion from Tangney in just a few months. How could I take care of Polly and be sent all over the world at the same time? I couldn't.

By the time I got to Shacknow's office, Rice had given him a heads up. He stood up from behind his desk and took a seat at the small conference table that sat in the corner of his office. He pulled out a chair next to his and motioned for me to sit.

I started to tell him what we'd found out. Before I could get the words out, I choked, then gulped, and finally just broke down and cried like I'd never cried in my life. Shacknow sat next to me not saying a thing, but I could feel his gentle presence. Finally, he said, "Tom, what can we do to help?"

I pulled myself together, and shaking my head I told him that I thought I'd have to turn down battalion command. I couldn't be there for Polly and be there for a battalion at the same time. It wouldn't be fair to either.

Shacknow then looked me in the eye and told me that he would support anything I decided to do, but I was not to make any decisions for at least one month.

I nodded in agreement and asked that I be able to stay as the Deputy G1 rather than attend language school. There was no way I could take on French now or ever for that matter.

Shacknow smiled and told me he'd work it out with MG Suddath. And he did. I will forever be grateful for the kindness that all in my chain of command showed me during that time. Most of all I will never forget how Shacknow treated me. He was a true gentleman as well as an outstanding soldier.

A COUPLE OF months latter Polly's symptoms began to clear. MS is like that. You never know when an attack will come, or what, if any, permanent damage it will cause. It is a fickle and cruel disease. We would learn to live with it. At the time of writing this memoir, Polly is doing great. But who knows what lies ahead.

I tried out for the team and was selected to represent Fort Bragg and the Army at the National Fitness Classic V.

Command 3rd Battalion 5th SFGA

MY TWO YEARS as battalion commander would also be the best of times and the worst of times.

April is a fickle month in Fayetteville, North Carolina. It could be hot, cold, or somewhere in between. On the 18th of April in 1987, the month smiled on us providing a mild, clear day. The change of command between Tangney and me went well. Bill gave an eloquent goodbye, and I gave a short hello. It was good to follow my best friend into command at Fort Bragg. Polly and I said goodbye to Bill and Kanti. The Army had assigned Bill to MILPERCEN in Washington, D. C.. He would eventually become the first branch chief for the Special Forces Branch.

The first thing I did was to gather my XO, Sergeant Major, and the three company commanders in the conference room and run through my command philosophy with them. That philosophy encompassed the Five Principles of Good Management (which I had renamed the Five Pillars of my Command Philosophy) as its foundation. I further stressed that they were never to lie to me. If I asked a question they could request that they not answer it. I would then decide whether or not I really wanted to know the answer. If I did, they would answer it truthfully. Often the coverup brought more heat than the infraction.

I also had to break them of the habit of jumping up and coming to attention when I entered a meeting. Shacknow never let us do that, and this humble display impressed me from the beginning. I swapped offices with my XO, MAJ Tom Chranko. He moved out of his little office and into my larger one that housed a long table with chairs. He could now hold his staff meetings without having to move to another building. I got this from my old battalion commander, Dan Edwards. He did the same for me when I served as his XO.

The Army had recently instigated a regulation which required that all NCOs and Officers receive an initial and quarterly written counseling from their rater. This was coincidently in line with one of the five principles/pillars I preached. After the meeting, I had a one-on-one with each of those I rated and gave them their initial counseling in writing, again emphasizing the five pillars. I further directed that they would do the same for all under their charge. Had they already done so, they were to add an addendum to their counseling forms explaining the five pillars and how they applied to the unit today.

I gave raters two weeks; then I visited each and asked to see their counseling. Most had complied with my request. I asked those who hadn't to do so immediately. As my old Boss Dan Edwards would say, "Those things get done only what the boss checks." And he was right.

AT FORT BRAGG and other posts throughout the Army, pay day was a half day. The units spent the first half of the day doing various things. Many had inspections and so on. In SF, not the case. I used our half day to meet with all soldiers in the battalion to have a question-and-answer session. The rules were simple. They could ask anything of me they wanted. They had to be respectful. I would not get upset with any questions and would answer as best I could. I particularly wanted to get questions from them concerning what they had "heard" that the Old Man had said or was requiring all to do that was "stupid." Often by the time a directive filtered down from me to the individual soldiers, the intent as well as the letter of the requirement became garbled. I could get direct and immediate feedback from the troops via these monthly meetings.

Sure enough at this initial meeting, someone asked why I was forcing the chain of command to give everyone written counseling. It was obvious that a breakdown had occurred concerning the Army's policy on counseling. I explained that while it was, in part, one of the five pillars of my command philosophy, it was also an Army regulation. All must receive initial and quarterly counseling.

They all got it. In the future, we would work through many problems at these monthly meetings. I got honest and unfiltered feedback directly from my soldiers. Sometimes they would catch me doing something stupid, and I would reconsider and take action to fix the problem(s) I created. Other times, I'd find out that a commander between me and the individual soldier had done something stupid, and I could address that as well later on. The rule that no one would be put down for expressing an opinion during one of these meetings was strictly adhered to by all in the chain of command. This was the key to making these sessions work.

I REGULARLY VISITED the A Teams' training. I tended to gravitate toward the SCUBA teams and would often dive with them. I would throw out challenges with the loser buying a case of beer and all drinking it. The first challenge I made was to ODA 85, one of my

SCUBA Teams. They were conducting their requalification dive training off of Onslow Beach at Camp LeJeune, North Carolina.

One of the things the Team had to do was an open water 2000-meter swim. The week before, I had tested my time swimming 500 meters with and without fins. I was fast with fins but even faster with a crawl stroke and no fins. I reasoned that few if any in the battalion could beat me at a distance of 2000 yards using only fins and me the crawl. My offer to 85 was that if any member of the team could come in ahead of me, I'd buy the beer. SFC Will Hayns immediately jumped on the bet, committing the rest of the team. Chief Warrant Officer Vericant, the current Team Leader, wasn't so sure, but went along with the contest.

We used two seven-person rubber Zodiac boats powered by forty horsepower Mars outboard motors to take us to the 2000-meter marker. We all rolled off the sides, leaving one person in each boat so serve as safety and follow us, one boat in the lead, and one boat pulling up the rear. The end results found me standing dry on the beach as SFC Haynes, the first in, walked out of the surf. We had our photo taken as a group. I stood in the middle with my red Triathlon swim cap on, goggles around my neck. Haynes stood to my immediate right, covering his face. Vericant, standing to Haynes' right, couldn't help but laugh.

MY S3, MAJ AL AYCOCK, had been pushing me to conduct a battalion level FTX. Al proved to be one of the most talented officers I'd ever worked with, and he would retire as a general officer. He was rigidly respectful to me. I made a mental note to loosen him up. I gave him the go, and he began putting the operation together. In the end, the battalion would set up as a Forward Operating Base at Fort Bragg and deploy a company headquarters and its teams to Fort Stuart, Georgia, to conduct a series of Direct Action missions. The company would operate in a simulated safe area on Stuart while the teams infiltrated their operational areas via chopper, parachute, and surface swim.

COL Harley Davis, who commanded the battalion I currently did when I was the Group S1, now commanded the 5th Group. He attended a couple of Teams' briefbacks. Afterward he told me he was impressed with the overall attitude of the soldiers. That meant a lot coming from him.

One of the briefbacks he didn't attend, but I did, was given by ODA 75, a newly formed SCUBA team in C Company. The Team had about only half its men SCUBA qualified, but all had been surface swimming with fins and equipment. They would infiltrate via a short 1000-meter surface swim. Coincidently, the Team Sergeant on 75 was my old swim buddy from SCUBA School MSG Erickson.

After the briefback, the Team showed me how they would secure their personal and team equipment during the swim. The only doctrine dealing with securing equipment during a surface or subsurface swim was that only safety equipment would be secured to the swimmer without a quick release, that safety equipment being the UDT vest (a life vest inflated via a CO2 cartridge in an emergency) and the dive knife.

THE FTX KICKED off in the second week of December and was scheduled to end just before the Christmas holidays began. ODA 85, a seasoned SCUBA Team, would serve as safety and support for ODA 75's night swim infiltration. All went well until 75 began the swim in. At some point, WO Pelver, the Team's XO, lost his M16.

The company reported the loss to me at the FOB. The swim was just off the coast south of Savannah, Georgia, near one of the inlets. ODA 85 had marked the spot with a buoy as best they could. I instructed the company to have 75 continue with the mission, reasoning that 85 would conduct a search for the weapon at first light the next day.

Losing a weapon in the military is a *very* big deal. Normally, the exercise would come to an immediate halt while the unit who lost the weapon would search the entire area until it was found. Of course, this would be when a weapon was lost on land en route to an objective. I saw no reason to do this under the circumstances. We reported the loss and my decision not to stop ODA 75's movement to the objective to the 5th Group's duty officer.

EARLY THE NEXT morning I packed up my wet suit and fins and drove south down I-95 to Fort Stuart. I linked up with MAJ Blue Keller, the company commander. We had choppers on standby to support the various Team exfiltration and infiltrations. I pulled on my wet suit and had a chopper fly me out to the area where Pelver had lost his weapon. ODA 85 had set up on the beach near the area and had its Zodiacs with divers searching the ocean for the lost weapon.

The chopper made one pass over the Zodiacs, and on the second pass, I did a helocast near one of the rubber boats. I finned my way to it. When I got there, I commandeered a set of tanks with regulator, weight belt, and UDT vest and dove down to see the conditions for myself. The river current that flowed just to the left of the dive site had swept the ocean floor clean. It was obvious to me that the Atlantic had long since claimed the weapon that had been wrapped in a protective plastic covering.

Even so, I had ODA 85 continue the search. I stayed with the company headquarters located on Fort Stuart the rest of the day and that night. The Group Headquarters passed the word that COL Davis and his XO, my old friend from Bad Tolz, LTC Vaden Bessent, were driving down.

When they arrived, I had 85 ferry us out in a Zodiac to the dive site. I explained to Davis what we had done in an attempt to recover the weapon and why I decided to have 75 continue with the mission, letting 85 conduct the search. Basically, ODA 75 only had a few SCUBA qualified guys, and they had not worked together long enough to conduct a search under the conditions we faced.

Davis seemed to understand. I told him we were calling off the search and packing up to return to Bragg. Christmas holidays were only a few days away, and I wanted to get back to Bragg so that those with leave planned could head home.

Early that afternoon Davis left with Bessent, and we began to pack for the redeployment. The phone rang. It was the Group's Deputy Commanding Officer, LTC Jeff Fuller. "I just got a call from Colonel Davis. He's en route back here and stopped to call. Message is for you to have the Team that lost the weapon continue to search for it."

This made no sense. In addition to the hopelessness of the search, we had a storm brewing off the coast that would hit the next day. No way would a dive master authorize anyone in the water under these conditions. I explained this to Fuller. He agreed, but said that he couldn't get in touch with Davis. This was before everyone and their brother had cell phones. Davis had called from a pay phone somewhere off I-95.

I hung up and noodled the problem around in my head. ODA 75 couldn't safely conduct the search for the weapon even if conditions were perfect. The weapon was long gone out to sea. I'd have to cancel ODA 85 and 75's leave plans to begin the search after the storm had passed. This could be two to three days away. In the end, I decided that the only option I had was to call the Chief of Staff, my old boss, COL Sid Shacknow, and explain the situation to him. I did.

Shacknow told me to wait, and he'd discuss it with the CG. MG Suddath, the 1st SOCOM commander, was raised in Savanna, Georgia, and knew the area well. Within the hour, Shacknow called back and told me that the CG agreed with me and to come home.

I HAD JUMPED the chain of command. A big NO, NO in the Army. It seemed like a good idea at the time, but proved to be one of the worst I'd ever have. No question about it. Looking back, I screwed up. When Davis returned to Bragg and found out what I'd done, he was royally and justifiably pissed off with me. I've often wished I could go back and undo what I did. In retrospect, I should have waited until Davis got back to Bragg and talked to him personally. I could have laid out the facts to him as I saw them, and surely he would have agreed to the redeployment.

When we got back to Bragg, Davis called me into his office and proceeded to chew on my ass. Something I couldn't fault him for. He ended by giving me a written letter of admonition for my actions. What could I say. He was right. I was wrong. I nodded, turned, and walked out of his office.

Whenever a weapon or any other sensitive material is lost, the Army reg requires that a 15-6 investigation be conducted. I explained to the investigating officer that the Team did all they could to prevent the loss. They had briefed me on how they would secure their equipment during the swim, and I'd approved. I also referred him to the SF Underwater Operations school in Key West. He confirmed with them that there was no set way to secure equipment to an individual swimmer. Their only guideline was that only safety equipment would be secured without a quick release. In the end, the investigating officer recommended that no one be held liable for the lost weapon.

As was his prerogative, COL Davis did not accept the investigating officer's recommendation and directed WO Pelver to pay for the weapon. This would cost the soldier almost $600.00 which was quite a bit in those days, especially for a warrant officer to cough up.

I informed Pelver of Davis' decision and of his right to appeal to the 1st SOCOM CG. I told him that I would support an appeal if he decided to do that. He did. The appeal had to go through COL Davis to MG Suddath. I endorsed it, stating the facts as I knew them and concluded by asking that if the CG determined that Pelver had to pay for the weapon, that I be held accountable for half the cost.

I don't know what Davis wrote in his indorsement. He may have finally agreed with me, but at any rate, the CG decided not to hold Pelver liable for payment.

This incident would somewhat taint my relationship with COL Davis. This was something that did not have to happen. Had I just given a little more thought as to how I'd handle the situation. . ... But as always, I had to do it my way.

SOON THINGS WERE back to normal, and I was up to my old trick of challenging the Teams. I had a policy that once a month the Teams would conduct a six-mile run. The goal for all soldiers my age or younger was to complete the run in under forty-eight minutes. Any soldiers older than me just had to complete the run. I used the word "goal" rather than "standard" because if a standard wasn't met, then I would have to take action against those who couldn't meet it. I did require that the times be posted in the Team room for all to see. This would increase peer pressure on all to meet and exceed the goal. Generally, all in the battalion would conduct the run on the second Monday of each month. Any who weren't there for the Battalion's monthly run had to make it up sometime during the month.

After the first two monthly runs, I offered a challenge to the Teams. Any Team who could put four runners ahead of me would win a case of beer. If a Team accepted the challenge and failed to put four in ahead of me, the Team bought the beer, and we'd all drink it at the end of the day.

ODA 95 (SCUBA) was the first to jump on the bet. I was a little worried as 95 had some very fast runners. SFC Putnam, one of my favorite soldiers, was one of the fastest runners in the Battalion. With his wide shoulders and chest and long legs, he was built like a combination of a weight lifter and a runner. As the run neared the end, 95 had two well ahead of me. I could see the back of a third runner in the near distance. A fourth Team member was closing in on me as we neared the finish line. I kicked it hard and just made in ahead of their fourth man. This was a close call for me, but I let the bet stand for the rest of the year.

THE BATTALION HAD been tagged to support this year's Flintlock exercise with a company in Zaire, Africa. The country was first known as The Belgium Congo. In 1997 Zaire changed its name to Democratic Republic of the Congo. The company and other of my units fell under operational control or OPCON of SOCEUR (Special Operations Command Europe) out of Stuttgart, Germany. My Battalion Headquarters had no responsibility except to provide forces to SOCEUR.

Company C under command of Blue Keller (Keller spoke fluent French) was tasked to deploy into Zaire and conduct joint training with Zaire's only airborne brigade. The French influence was still strong in Africa south of the Sahara. In peacetime, a French Colonel commanded the Zairean airborne brigade with a Zairean Colonel as his deputy. In times of war the positions would reverse.

As the battalion commander, I was scheduled to visit the training going on in Zaire. Keller met me at the airport near the capital city of Kinshasa. It was typical third world–small and unorganized–but I finally made it through customs.

"You didn't pack an alarm clock or anything like that in your baggage?" Keller said when I left the customs area.

"No. I got the word you passed on." Nothing of any value made it through the baggage handling.

We rode in a military vehicle from the airport to a small army base. A civilian rental wouldn't work here as we would surely be stopped along the way and have to pay bribes to the local police. Zaire couldn't function without the corrupt system that had been fostered by Mobutu, the country's current president for life. The pay to government officials was so little that graft and bribes were actually part of the system.

I spent two weeks with the company. The Zairean and French Colonels had a party in honor of my arrival. The food was local, and I knew what would happen if I ate it. However, not to eat it would be a great insult. I probably shouldn't have eaten a little bit of everything on the table, but it tasted pretty good. I was hungry. I was also careful to only drink the bottled beer provided in great quantities. That along with a couple of shots of whiskey I figured would kill any little things that might attack my intestines. Wrong. Sure enough within three days, my stomach cramped with diarrhea. It wasn't as bad as Vietnam, but it was worse than my trip with Tommy Tucker back from Mexico City to Monterey, Mexico.

Keller had set me up with a room in an air-conditioned building that housed his company headquarters. SFC Putnam also had diarrhea, so I moved him in with me. The Teams endured the oppressive heat and lived in what were basically mud huts. Putnam and I would suffer together, but we wouldn't be sweating our asses off. Our PA, Mike Davidson, finally arrived with some little green and white capsules that did the trick.

C Company had brought along an assortment of special staff to assist in their Foreign Internal Defense mission. They ran a civic action mission as well as training in tactics, weapons, and airborne operations. As part of their Civic Action mission, CPT Rich Barbaro, the Group's dentist, was there to provide dental care to the Zairean soldiers and their family members. I will never forget seeing Rich set up under a large tent sheltered by palm trees, a big smile on his face, covered from head to toe in protective clothes, including a pull-down plastic face mask (AIDS was rampant in this country), drilling and/or extracting the teeth of any who showed up. All the while a portable generator whined in the background. Rich was a soldier first and dentist second, and great at both.

WE HAD A wing exchange jump scheduled. We would award the Zairean paratroopers our wings, and they would reciprocate by awarding us theirs. They would jump from a US C-130, and we would jump from one of their C-130s. It was an experience I'd not soon forget.

We loaded into the Zairean C-130 to find that it had no web seats. Instead it only had plywood on the floor with one long safety strap that went from the front to the rear on each side of the aircraft. In the forward portion, tires lay haphazardly secured by straps. My guess was that they were being transported somewhere after we made our jump.

We were also required to jump their parachutes. That wouldn't have been so bad, except I could see pieces of the chute peaking out between the seams of the backpack. Something that didn't instill confidence in me. We did insist on the reserve being our US packed ones.

The Zairean jump master used the same commands as we did. I made sure that I stood in the door first, as getting out of the aircraft was foremost on my mind. As I stood there looking down at the palm trees and high grass race below, I could see an occasional antelope dart left and right, frightened by the aircraft as it passed overhead. The Zairean jump master gave the command to "GO," and I held a tight body position as the turbulence whipped my feet up and above my head. I felt the familiar tug and heard the pop of opening nylon. I was never so glad to look up and see an OD canopy over head.

The company set up its field training headquarters in an area near the Angolan border. Kamina Base was at one time a Belgium Army installation. It basically sat in the middle of nowhere, but had adjacent to it an air strip that could easily handle a C-141 and probably a C-5A jet aircraft. There was no logical reason for an airstrip of this quality there. However, at night we could hear light aircraft taking off, heading in the direction of Angola. I suspected that they were CIA in origin, and the heavy duty airstrip was there as a contingency plan for activities that might have needed to take place should an invasion of Angola be necessary. This plan was obviously overtaken by events.

Near where the company set up stood an old Belgium Officers' Club. Better described as a palace. A majestic spiral staircase led up to the second story ballroom that must have been a good 10,000 square feet. Windows twenty-feet high and seven- or eight-feet wide looked out over two Olympic size swimming pools. French chateaux surrounded the majestic building.

That was what it must have looked like when the Belgians occupied the base.

Now the doors were torn off. Goat scat covered the spiral staircase and up onto the ballroom's floor. The large windows were ripped out. The once grand swimming pools now had trees growing in the cracks that scarred their empty bottoms. Everywhere that electrical wire had run were large gaping holes where the wire had been torn out for its copper. The chateaux were also gutted, bearing only a slight resemblance to the grand structures they must have been. All stood as a sad reminder of times gone by.

No deployment of this type is without its dangers. C Company had scheduled a mass tactical jump with one of the airborne battalions. US C-130s landed on the airstrip and loaded both Zairean troops and members of C Company. I watched from where the DZSO (Drop Zone Safety Officer) had set up.

The third aircraft in the formation emptied its jumpers. About midway of the stick, two Zairean jumpers tangled with each other. The chutes alternately stole air from one another as the two leap frogged down. I knew it would be bad, and it was. One of the soldiers died on impact. The other had to be evacuated to the Brigade's base near Kinshasa. This accident would stain what had otherwise been a successful deployment.

I left a few days after the accident and caught a civilian flight up to Tunisia where I had a unit training with the Tunisian soldiers outside Tunis. After a few days there, I flew into Pisa, Italy, to observe training of another of my units. Leaving from the Rome Airport, I winged my way into Heathrow Airport in London. Between the tube and the train, I finally ended up at RAF Sculthorpe, where SOCEUR had set up its command and control headquarters. I discussed the fatality we had had in Zaire with BG Scott, the SOCEUR commander. Three days later, I caught a flight back out of Heathrow and finally returned to Bragg and Polly. I guess you can now see why I'm not a real big fan of flying, specially when that flying crosses multiple time zones.

AT THAT TIME every Battalion had a FSG (Family Support Group, later designated Family Readiness Groups). The wives of the battalion made up the group, and it was normally headed by the Battalion Commander's wife. The wives banded together to take care of one another when the Battalion was deployed or back in garrison and any emergency occurred within the unit. They also served to assist newly assigned soldiers' families in their move into the area. These groups generated money for their activities by having bake sales and such.

I had the idea of the SFG sponsoring sporting events—Triathlons, which combined swimming, biking, and running, and Biathlons, which combined running and biking. Actually, the wives would handle the small stuff while the Battalion's soldiers would handle the bulk of supporting the event and participating in it. In that regard the Show Me What Y'all Got series of events began. The Battalion hosted the events, and all profits went to the FSG. As a result, the FSG's bank account swelled to the point that we were afraid that we were in violation of the Army's regulation dealing with private organizations. The bottom line was that not one cookie or cake ever had to be made or sold by any of the wives in the Battalion's FSG.

IN JANUARY OF 1988, Bob Boeder, a friend of mine and ultra marathon runner, came up with the idea of a fifty-mile race for individuals and teams. He asked if I'd be willing to sponsor it on post while members of the Fayetteville Running Club would provide all support. I agreed.

The battalion would enter a five-man team. Each man would run ten miles in relay. As a result of our six-mile monthly runs, I knew who my top five runners were. The day of the race we had had snow mixed with ice, and the roads were dotted with slick spots. Boeder didn't want to call the race off as he had well over one hundred participants signed up.

I was there not as a runner but as support to the battalion team I'd put together. Putnam, from ODA 95, was my lead runner. I told him to check the course out and, if he thought it too dangerous, to let me know when he came in. I'd then cancel the team's participation.

I stood waiting for Putnam to near the start/finish line. When he came into sight, I ran along with him and asked if he thought I could call off the run. He shook his head "no," and tagged off the next runner.

We took first place in the team competition, and each runner received an individual plaque.

At the next Battalion's monthly meeting where all got a chance to ask me questions, Putnam had the balls to stand up and say, "Sir, I can't believe that you had us run that fifty miles in all that ice and snow."

As I've written previously, Putnam was one of my favorite soldiers. I smiled, then said, "All you guys know that you can ask anything in these meetings, and I won't criticize or jump in your shit about it. But I'm going to make an exception this one time for Putnam." Putnam had sat down by now, and I motioned for him to stand back up. "Okay, Putnam, I want you to tell everybody what I told you before you started the race."

Putnam knew he had stepped in it. He hesitated, not wanting to answer the question.

"Didn't I tell you I wanted you to check out the course, and if it was too dangerous, I'd cancel the rest of the runners?" I was fighting back a laugh. "And when you came in from the run, what did you say when I asked if we should continue?"

Putnam gave a sheepish nod, sat back down, and all burst out laughing.

EVER TRYING TO come up with a challenge for the Teams, I proposed that I could swim farther underwater with one breath than anyone on any Team. The SCUBA teams couldn't pass up on that. My only requirement was that I'd pick the place and the uniform for the swim. Both 85 and 95 were chomping at the bit to ream me out, and this, they thought, would be their time.

I had perfected a technique of swimming underwater without using my arms. I had a strong frog kick that would send me through the water just as fast and as far as most swimmers could swim using both arms and legs.

The Lee Field House pool had lap lines that were raised tile. I had practiced swimming along the bottom with my hands out front feeling the tiles and knew I could make the twenty-five yards down and back with no problem.

I told 85 and 95 to put their best together and meet me at the Lee Field House pool for the swim. When all arrived that morning, I passed out blacked out face masks.

They couldn't believe it.

"What the hell is this for?" SFC Haynes said.

I eased into the shallow end of the pool. "Remember that I said I would prescribe the uniform? Well, blacked out face masks are part of the uniform. Let's get it on."

They begrudgingly got in their lap lanes and pulled on the masks. It was a hoot. I dropped down and felt my way along the raised tiles the entire length of the pool and back. Not being able to see where they were as they swam, some ran into each other. Others swam until they thought they were about to reach the end of the first lap then tried to feel out in front, fearing they would crash head on into the side. Bottom line was that none made it all the way down and back.

"You tricked us again," was the cry of the day.

After the underwater swim, Haynes challenged me to a breath holding contest. I was reluctant, but accepted the challenge. He beat me. Well, I had to give them something.

I HAD YET to lose a Team challenge, and they were beginning to back off taking the bait. I had to come up with something that seemingly I couldn't possibly win. I proposed a weight-lifting contest. The rules were that we'd see who could hold a ten-pound weight over their heads the longest. Whoever let the weight dropped below shoulder level first would be the loser. I then added that I would use my strongest arm and the challenger would have to use his weakest arm. I also dictated that I'd pick the time and place of the contest. Well, how could they lose this one? They focused on the strong arm vs. weakest arm requirement, and not the fact I'd be picking the place.

ODA 85 had been bit too many times and wasn't going to bite again. ODA 95 had on their Team a guy who was a weight lifter. I don't remember his name, but his arms were bigger than my legs. They knew they had me now.

I told them that we'd meet in the weight room of Lee Field House at 0800 the following morning. That morning all of 95 and several from 85 gathered in the weight room, sure they were going to see me go down in flames.

I picked up a ten pound barbell and handed one to my opponent. He took it in his left hand and held it up over his head, saying that he'd get started now and that I could wait five minutes before lifting mine.

I allowed as how that was fine with me, but this wasn't where the contest would take place. I then turned and motioned for all to follow me. I walked through the dressing room and into the pool area. Now the Team knew what was about to happen, but couldn't do anything about it. I jumped into the deep end and motioned for my opponent to follow. He did.

I had always been able to tread water for as long as I needed, and holding ten pounds over my head was a snap. This wasn't so for the sergeant who had taken the bet. He lasted about ten minutes then began to bob with his head going under then back up. To his credit he struggled far longer than I'd ever thought anyone would, but in the end he went down. I'd beat 'em again.

MY NEXT CHALLENGE was to put myself against a Team's fastest three swimmers in a 1500 meter race. Each of the three would swim 500 meters while I'd swim the entire 1500 meters. Several Teams took me up on that one. On any one team there might be one who could swim faster than me, but that guy would only get to swim 500 meters. There was no way that the other two would be able to maintain the lead. I'd skunked them again.

ODA 85 wanted to try the relay race but they wanted to use fins and arms while I'd only use my arms. I knew that their top three swimmers would beat me, but agreed anyway. I was holding my own through the first swimmer, but began losing ground against the second swimmer. The third maintained the lead and added a few yards.

They beat me by about fifty yards, but when I finished, I said, "Okay, you guys beat me, but if you want bragging rights, you have to do it again. Now."

They all begged off. I was relieved.

THE ASSAULT ON Mount Mitchell is a bike ride/race 102 miles (164 km) and has over 11,000 vertical feet of climbing. It starts in Spartanburg, South Carolina, and ends at the top of Mount Mitchell in North Carolina. Forever looking for a challenge, I convinced my lawyer brother, John, that we should give it a shot. John had not only been running but also biking ever since I roped him into doing the Penrod's Triathlon.

John and I completed the 1988 Assault in somewhere around ten or eleven hours. I have to tell you that it was a painful experience. It took us longer to peddle the last thirty miles than it did to peddle the first seventy-two. I am pleased to write that I made it all the way without having to get off my bike to walk.

The next year John convinced two of his lawyer buddies, Bobby Chasteen, an attorney and now a Superior Court Judge, and Mallon Faircloth, a Superior Court Judge who has since retired as a U. S. Magistrate, to join him. I tricked my friend and Triathlon buddy, Chuck Brisco, into going as well. Again, I was the only one who stayed in the saddle for the entire 102 miles. John and I decided that the first time we did it out of ignorance. The second time we did it because we were stupid. But I must say that it provided all with some great "Been there, done that" stories.

I'D HAD ABOUT a year in command when I heard that the SF School (SWC) would be going back to the Training Group concept. This would mean that a full Colonel would be the Training Group Commander. He'd serve under SWC's two-star general. Under the training group would be three battalions commanded by lieutenant colonels. My old boss from my time as 5th Group's S1, MG Guest, was the commander of SWC. Told you SF was a small world.

I made an appointment with him. I wanted to ask him to let me stand up the Second Battalion. This unit would have four companies under it. What was now the committees responsible for training HALO (High Altitude Low Opening Parachuting), UWO (Underwater Operations), O&I (Operations and Intelligence), SERE (Survival, Evasion, Resistance, and Escape), and SOT (Special Operations Training or close quarters combat) would be designated as companies and fall under command of the 2nd Battalion.

Guest agreed but wanted to know what would happen if I came out on the War College list. The Army War College is the highest schooling an officer can attend. Competition for selection is fierce. To be selected to attend as a combat arms officer, you had to have first been selected for and successfully command a battalion. If selected, a Lieutenant Colonel was pretty much assured of making full Colonel. I thought I had a chance of being selected, but didn't bet on it.

I told Guest that if selected, I'd turn the resident course down and do it by correspondence. He thought about it a minute, then agreed, and we sealed the deal.

Sure enough, I popped on the list first time I was considered. I was both surprised and amazed. I called MILPERCEN and asked the officer that managed the War College list to pull me off the resident list and put me on the correspondence list. The guy tried to talk me out of it, but in the end said he's see what he could do.

By this time, the 5th Group Headquarters and its first and second battalions had moved from Fort Bragg to Fort Campbell, Kentucky, leaving my battalion behind. We would eventually become the first battalion of the 3rd Group. I called COL Davis the Group commander and told him what I'd done. He thought I was crazy and wasn't at all happy I had turned the down the school. I explained to him the deal I'd made with Guest. He offered to call Guest and get me out of the commitment. I refused. As always, I'd do it my way.

MY COMMAND OF the 3rd of the 5th at Fort Bragg was almost at its end, when at 2330, on the 12th of March, 1989, I received a call which I always knew could come at any time but hoped and prayed never would.

The duty NCO said, "Sir, we have just received a call from Chief Britton from C Company. He said that two helicopters en route to Gila Bend, Arizona, have crashed and twenty-three of our people have been killed."

Company C had deployed with its new commander (of only two weeks) to Fort Huachcua to conduct desert environmental training. I couldn't, because I didn't want to, believe it. After questioning the NCO on duty who had received the call, I told him to get back with Chief Warrant Officer Britton, give him my home number, and tell him to call me ASAP. After several minutes, which seemed like hours, the phone rang, and on the other end I heard a voice which I recognized only too well as Chip Britton's. He told me that it had not been two helicopters but one which had crashed and that the manifest listed eleven of our soldiers as being on board. I had him read off the names to me as I numbly wrote them down.

At the time, there were twelve active and twelve National Guard Special Forces Lieutenant Colonels commanding the twenty-four Special Forces line battalions. What had just happened constantly hovers in the realm of possibility and gnaws at the guts of each one of us. Our collective missions distributed our 216 separate maneuver units over the entire world and placed us in close proximity to a tragedy of this nature daily. Our wartime and peacetime missions were the most personally demanding and sensitive of any of the US Army's forces. Because of this, we had to constantly participate in realistic and demanding training. However, even though circumstances repeatedly exposed us to this clear and present danger, none of us was really ready to deal with a tragedy of this scope. What do you do when you are in a command position and find yourself responsible for dealing with something of this magnitude? I must admit that on that night, as I sat staring at the eleven names of those fine soldiers and thinking about their families, I had no idea. The one thing I did know was that something had to be done immediately.

I called the Staff Duty NCO back up and gave him the correct information. I then told him to immediately notify the EOC (Emergency Operations Center) of the 1st Special Operations Command (SOCOM). After this I called my Battalion Executive Officer and my Command Sergeant Major. My next call was to COL Davis at Fort Campbell, Kentucky. MAJ Blue Keller had just turned over the company and was currently assigned to SWC. Since he and his wife, Judy, knew all the families, I called him. He and Judy proved invaluable in helping us get through the next month. By this time it was 0030 on the 13th of March. I then called my XO back and told him to contact all the senior individuals left back at Fort Bragg from C Company and any other senior personnel from within the battalion. The plan was to assemble at 0500 hours that morning wearing our dress green uniforms. These individuals would report to the Battalion conference room. I knew that we would have to provide initial notification to the families, and I wanted to make sure that we had at least two and, if possible, three individuals per family.

The Battalion's S1 officer immediately drove in and pulled the deployment packets each man had filled out prior to departing. Those packets contained a copy of an up-to-date emergency data card that we'd made for each man prior to his departure. On it was all the information we would need to determine who was to be notified in case of an emergency. The S1 also acquired city maps and began to mark the location of the homes of the men involved. It was here that we encountered our first major problem. Some of the men lived well outside of the city, and their addresses were not on the map. Also, in some cases the maps we used proved inaccurate or misleading. It would have been of great help if we had had all the men who deployed draw a sketch map to their homes as well as where their wives' worked.

The wait from the time of that initial phone call until 0500 that morning seemed like an eternity to me, but it was nothing compared to what was about to begin for the unit and the soldiers' families. At that early morning meeting, I passed out all the information that we had and told the officers and senior NCOs present what we planned to do. First, we had to receive a briefing from the Fort Bragg Casualty Assistance

Office. Next, we had to give the initial notification to the families involved. I wanted very much to receive that briefing immediately so that we could get to the different homes before the wives left for work. We would then meet back at the conference room to provide an update on the situation.

The Fort Bragg casualty office wasn't prepared to deal with something of this magnitude either. Even so, most of us made it to the families by 0800. Each situation was different, but all were traumatic for both the families and the men of the battalion who had the unenviable task of delivering the news. It hurts even today as I vividly remember the reaction of the wife that I and two other officers notified. All we could tell her about the accident was that her husband's name had appeared on the manifest of a helicopter that had crashed in Arizona and that the incident was under investigation to determine who was actually involved. This initial notification gave some glimmer of hope that there may have been a mistake and that her husband was either not on board or had somehow survived the crash. And at that time this possibility may have existed.

I had instructed the men that they were not to leave the families alone after the notification and that we would ask other wives in the FSG (Family Support Group) who knew these families to come and stay with them until their relatives or friends could arrive. Since at least two men were involved in each notification, one could stay with the family while the other could return to the battalion area to provide an update. At this point, I would be remiss if I didn't recognize the wives within the battalion who rallied to help during our time of need. Anyone who might ever sell the Army wife short should see them work together in a situation like this. Their support and dedication were nothing short of miraculous.

Polly and I had had loud discussions in the past, and we would have loud discussions in the future, but when I needed her most she was there for me standing tall and strong. I will forever be grateful to her for this.

While we, the soldiers, and leadership within the battalion dealt with this tragic and fluid situation, the wives who composed the FSG were tending to the immediate needs of the families. The FSG offered everything from meals to house cleaning and baby sitting. Soon the full support of the entire military and civilian community flowed our way. Fayetteville Technical Community College, where Polly taught, stepped up with both food and money solicited from the administration, faculty, and students. Other units within the Special Operations Command responded in like manner. The burden of ensuring that this generous outpouring of support reached the families in an appropriate amount and a timely manner rested on Polly and Judy Keller, and those many other wives who composed the family support network.

The crash was what you'd expect under these circumstances. The aviation fuel combined with the magnesium in the skin of the aircraft burned for several hours, making search and recovery operations extremely difficult. However, those individuals from the nearby Air Force Base, Davis-Monthan, worked around the clock in an effort to determine the status of all those on board. Everyone on the two installations involved, Fort Huachcua and Davis-Monthan AFB, provided every assistance to the remainder of the company as they prepared to redeploy.

Back at Fort Bragg we found ourselves operating under extreme pressure. I called for a Battalion formation as soon as possible on the morning of the 13th. At that formation, I informed, as best I could, the rest of my men what had happened and what we were doing to cope with the situation. I told them what we were allowed to tell the families and asked for their assistance in dealing with this crisis.

I had set up two briefings daily, one at 0900 and one at 1600 hours. All those who made the initial notification attended these briefings along with the individual chaplains which Fort Bragg now assigned to each of the families. During these sessions we received updates on the status of the different families and exchanged ideas as to how to best handle the multitude of situations we found ourselves in. We also received briefings as required from the installation casualty officer, social services, and personnel trained in the psychological aspects of what we were going through.

It soon became clear that all those whom we believed to be in the accident were, in fact, dead. The major problem at this point was that Positive Identification had not been made, and until this had happened, we couldn't give the families final notification. The news media had provided almost immediate coverage of the incident and had flatly stated that all involved were dead and, "names were being withheld pending notification of the next of kin." This, coupled with the fact that the Air Force had given final notification to the four families of their pilots and crew members, added to our frustration. Even though by 1200 hours on the 14th of March we had a 100% accounting for all remaining C company personnel and fifteen bodies had been recovered from the crash, we still could not get permission from Department of the Army's Casualty Branch to give final notification. By the morning of the 15th, most of the families had given up hope and had resolved themselves to the inevitable. The remainder of C Company returned that evening. I met them at Pope AFB and briefed them on what had been going on here at Fort Bragg. Approval for final notification came at 1130 hours on the 16th of March.

On Friday morning, the 17th of March, I gathered the entire battalion together and described in detail everything that had happened at Fort Huachuca. Questions were asked and answered to the best of our ability. Al Aycock, the Battalion S3, gave an accounting of our efforts back at Fort Bragg. Again, the men were encouraged to ask questions. Finally, I had the officers and NCOs who had been with the families since the beginning brief the battalion on the status of their assigned families. This was a very important meeting. It served to stop the rumors that were circulating and helped all of us better deal with this tragic event.

I made the decision to provide the CAOs (Casualty Assistance Officers) from within the battalion and selected them from those who had been involved with the families from the beginning. These individuals would be responsible for assisting the wives in attending to all the administrative actions involved with their husbands' deaths. It was a questionable call on my part since the individuals could be involved with these matters for quite some time. I elected to do it anyway. This gave me absolute control over the selection of the CAOs, and I could insure that they would receive maximum support from the Battalion while serving in that position. I also reasoned that the consolidation of these individuals from a single unit would provide a more efficient exchange of ideas, thus producing immediate feedback on lessons learned during the course of performing their duties. As it turned out, this proved to be one of the best decisions I made.

Polly had visited the families several times since the initial notification, but because of the situation I found myself in, I had not. So on the 18th and 19th of March, MAJ Blue Keller and his wife, Judy, Polly, and I visited all the families. It proved very difficult for me but not nearly as difficult as what I was to do the following day.

The memorial service was organized as a joint effort of the battalion and 1st SOCOM's chaplain section. It took place at 1100 hours on the 20th of March. I gave the commander's tribute to the families and the soldiers of the battalion. Mixed emotions stirred within me, but it was my responsibility. I had something that I wanted to say to all concerned. That something was this:

> "This is the hardest thing I have ever had to do. For the past two years it has been my privilege to serve as the commander of the finest soldiers in the Army. Today some of those soldiers are not with us. I know that there is nothing that I can say or do which will take away the hurt and fears you must bear. If there were, I would do it. We in Special Forces have always had a unique kinship with one another. So I would only ask that you please allow us to share in your grief. . . because we loved them, too."

ONE MONTH LATER, I turned the Battalion over to my old friend, LTC Frank Tony. He would do a great job. MAJ Andy Anderson, another old friend from my days as 5th Group S1, had just been assigned as a company commander. He was in charge of monitoring the funds held by the Family Support Group. Quite a bit had been expended in support of the tragedy we all had just experienced, but Andy told Frank that there was enough remaining so that not a cookie or cake would have to be baked or sold as the coffers were in good shape.

At the end of the 2000-meter swim, we had our photo taken as a group. I stood in the middle with my swim cap on, goggles around my neck. Haynes stood to my immediate right, covering his face. Vericant, standing to Haynes right, couldn't help but laugh.

I pulled on my wet suit and had a chopper fly me out to the area where the weapon was lost. ODA 85 had set up on the beach near the area and even had Zodiacs with divers searching the area for the lost weapon. I'm the one to your right.

The Most Fun I Ever Had With My Clothes On

ODA 95 six mile run challenge. I'm the second from the right, Putnam is the one just to my left in front of the guy in the Ollie North T-shirt. The soldier you see to the far right is the one who would lose the "weight lifting" contest to me.

The Battalion's 50 mile race five man team. SFC Putnam is in the center.

Assault on Mont Mitchell 1989. From left to right Mallon Faircloth, Bobby Chasteen, Me, Chuck Brisco, John Davis.

Command 2ⁿᵈ Battalion, Training Group

AFTER PASSING COMMAND of 3ʳᵈ of the 5ᵗʰ to Tony, I drove over to Smoke Bomb Hill, the original location of Special Forces at Bragg. The 2ⁿᵈ Battalion's headquarters sat off Son Tay Road in old wooden World War Two buildings. Waiting for me in the office was CSM Troy Graham. Graham was a bear of a man and black as midnight. Imbued with an abundance of common sense, he could state a problem in the fewest number of words as anyone I'd ever worked with. Loyal almost to a fault, he and I would work well together. I could always count on him to give me his unvarnished opinion, a trait I highly valued.

My XO, MAJ Palmer, was tall with a receding hair line and a whiz with computers. These machines were just becoming commonplace within the lower levels of SF and were a bit of a mystery to me. The computers were MSDOS based. This was well before Microsoft came out with the Windows Operating System. Never liking to depend on others and somehow knowing that these machines were the wave of the future, I had Palmer begin the unenviable task of educating me. I was particularly intrigued with the spell-check ability. The memory typewriters had this capability, but it was painful to use. I'd always believed that I had a learning disability when it came to spelling and was eager to master this new machine. Now, the problem would be picking the right word out of a long list offered when I murdered a spelling. Well, nothing's perfect.

My immediate boss, the Training Group Commander, was a full Colonel named Jack Moroney. Jack was built like a fire hydrant and just as hard. His bald head had a strip of hair that circled above his ears and rounded the back. He'd commanded at every level in SF and would prove to be the right guy to establish the Training Group. All of us would command our units until the Army and 1ˢᵗ SOCOM filled our positions via the centralized selection process. This would take twelve to eighteen months. At that time, we would be replaced by officers who had made the cut via a command selection board.

The move to reestablish the old Training Group structure stemmed partly from a desire to gain more command positions for both officers and NCOs. Command for officers was critical for advancement in the Army. The Training Group produced numerous commands at the company and battalion level for both officers and enlisted. It also served to establish another command position at the full colonel level. Additionally, it would provide a command structure for the soldiers/instructors whose job it was to teach other soldiers the skills necessary to become Special Forces qualified and also train them in the advanced skills necessary to perform their very sensitive duties. In other words, instead of being instructors first and soldiers second, they became soldiers first and instructors second: a subtle difference but an important one. Especially when the members of the Training Command rotated back to a regular Special Forces unit. At least that's the way I saw it.

THE BIGGEST PROBLEM Graham and I faced was effecting the paradigm shift from being a member of a committee or directorate to that of being part of a unit–Company and Battalion. The 2nd Battalion had four companies.

The HALO (High Attitude Low Opening) or Military Free Fall Company. This company provided military free-fall instruction for not only SF but any of the units in the Armed services who required this means of infiltration.

The Operations and Intelligence (O&I) Company. This company would train the O&I Sergeant (MOS 18F) who served on an A Team, giving him the knowledge and skills necessary to plan, organize, train, advise, assist, and supervise indigenous and allied personnel on collection and processing of intelligence information.

The SERE (Survival, Evasion, Resistance, and Escape) Company. This company would train Special Operations Forces in avoiding being captured and if captured, how to resist interrogation, and eventually how to escape and survive off the land.

The SOT (Special Operations Training) Company. These guys taught specially selected Special Forces soldiers in the art of close quarters combat, hostage rescue, and other skills akin to those employed by Delta Force.

The SCUBA or Underwater Operations Company. While the first three called Fort Bragg home, the SCUBA company sat on Fleming Key, a two-mile long island off the northwest corner of Key West, Florida, connected to Key West by a short bridge. You can guess which company I would determine needed the most of my attention.

As with 3/5, I gathered my company commanders and their sergeant majors together and talked them through the five pillars of my command philosophy. I asked them where they stood on initial and quarterly counseling of their soldiers. Blank looks all around. I gave them two weeks to get the counseling completed, and after the collective meeting, met with all that I rated and gave them their initial counseling.

I also held a monthly battalion meeting with all my soldiers. It was difficult to get all together at one time since their class schedules normally didn't stop for payday activities, but I did the best I could.

At the first of these meetings, an NCO stood up and expressed his displeasure at having been given an initial counseling statement and wondered why I didn't trust them to do their jobs without having it in writing. This was *déjà vu*, but by now the Army's counseling program had been in effect for over two years. See what I mean about the training committees needing a little push back into the real Army?

MG GUEST HAD moved up from commanding SWC to take command of 1st SOCOM, and MG Dave Barato moved into SWC. Barato was big into computers and required all to learn PowerPoint. That wasn't a problem for me as I had Palmer to set up all our slides. Eventually, I'd master this program as well as others.

One of the things that was killing soldiers who were trying to make it through the Special Forces training was the Land Navigation course. This was the subject of one of the Barato's command and staff meetings. Everyone had an opinion as to why the students kept failing the land nav course. I knew that it was one of the toughest courses around and thought that the guys who were throwing down on the students needed to get a taste of what our students faced. Most in the meeting hadn't picked up a map and compass in ten-plus years, much less wandered around Camp McCall at night trying to find a spot about ten-feet wide at the end of a 4000 meter dog-legged course.

I raised my hand, and Barato recognized me. "I recommend that we all go out to Camp McCall Saturday and run the day and night course. This will give us an idea of what the students are up against and what we might do regarding why so many fail the course."

That suggestion met first with silence, then some uneasy shuffling of asses in seats, then a clearing of throats, finally ending with Barato asking if anyone else had any other suggestions. Well, I thought it worth a shot.

IT DIDN'T TAKE me long to decide that I had to visit my company down in Key West. I called and spoke to MAJ Drake, the company commander. He would set me up with a BOQ room on the Navy annex. I told him that I'd like to throw out a challenge to any who'd take me up on it. I suggested that we conduct a six-mile race. I would designate the route and uniform. Did he think any would be interested? He said he's ask around.

I had loved Key West ever since I realized my childhood dream of becoming a Frogman by attending the Special Forces Underwater Operations course there in '72. If you haven't been there, you gotta go. The Conch Republic is like no other place in the States. People are so laid back that the prone position is considered a strenuous form of exercise.

I landed at Key West's small airport after changing planes in Miami. Drake–tall, dark haired, and lean as a scarecrow–picked me up. I checked into the Navy Annex's BOQ, rented a bicycle, and peddled the two miles out to the end of Fleming Key where the UWO School sat. Fleming Key is a long, finger-shaped island or key as they call it in the Keys. A short fifty-meter bridge connects it to the Navy Annex on Key West. A road runs down the middle of the Key and dead ends at the School. Most of the way along the road you can look right and left and see water. When I got to the school, Drake had four guys who stood itching to take me up on my challenge.

"Okay, guys. Here's the deal. The six-mile rout starts here at the school. We'll run down the road to the bridge, turn around, and run back here. Then off the pier and swim, without fins, one mile along the coast, turnaround at the buoy. Drake will drive one mile down and throw out the buoy. Then finish back here where we started. Any questions?"

It was a fun race and one guy almost beat me. Three of the four were well ahead of me when they hit the water, but I managed to catch up with two of them by the time I reached the buoy, and passed the last one only about 200 meters from the finish line. Anyway, it was a good way to introduce all to their new Battalion Commander.

I CAN'T SAY that much exciting happened during the eighteen months I commanded the 2nd Battalion. The training was some of the most high risk the Army offered, so safety was always a concern. Fortunately, no one was killed during my command tour.

Because of the vastly different training schedules in the Battalion, I left physical training up to the individuals. I did have a monthly four-mile run for all in my headquarters. That left the individual companies to work their own PT schedule as training permitted.

However, not willing to leave well enough alone, I dreamed up a two-day event that I dubbed The Gut Check. I required all officers in the Battalion to participate and invited any of the NCOs who wished to do so as well. The events in the Gut Check were based on performing a task against a specific time and challenged the individual to compete against himself rather than others.

The first of the two days started out at the Lee Field House's track and consisted of:

The number of pushups in two minutes followed by a three-minute break.
The number of bent leg sit-ups in two minutes followed by a three-minute break.
The distance run in sixteen minutes followed by a fifteen-minute break where all moved to the pool.
The distant swim in thirty minutes followed by a ten-minute break.
The distant run in ninety minutes.

End of first day.

The second day consisted of:

The distance walk/run carrying a forty-pound rucksack in six hours. This rucksack event circled a two-mile course, thus cutting back considerably on the support requirements.

I have to admit that all who participated did extremely well. We would conduct the event every six months. Would the fun ever stop?

MY TIME IN command of the battalion was drawing to a close when my name came out on the Colonel's list. I was now a Lieutenant Colonel Promotable. Of course, it would be almost two years before I'd pin on my rank. It was a great feeling, but it also meant that I would probably be reassigned from Bragg. At this point, my family had been living uninterrupted in Fayetteville for seven years. This was unheard of for an officer.

I got a call from the Special Forces Assignments Branch in MILPERCEN. They wanted to move me from Bragg. But where? We discussed options, and my assignments officer mentioned that there was an opening for the J3 of SOCK (Special Operations Command Korea). And just who do you think was commanding SOCK? None other than my old battalion commander, COL Dan Edwards. I couldn't believe my luck.

I could take a short (one year) unaccompanied tour with a home base back to Fort Bragg. This would mean the Polly and the kids could stay in Fayetteville while I went to Korea. At the end of my tour, I would be assigned back to Bragg for what could be another three years! Too good to be true. But before that could be locked down, I had to find a job for a full colonel on Bragg. As luck would have it, MG Sid Shacknow would soon be taking command of the Special Forces Command. I called him, and he agreed to make a slot for me even if it was temporary with the ultimate job being his Chief of Staff. At any rate this would give my assignments officer a legal way to send me back to Bragg after my year in Korea.

The assignments officer confirmed with Shacknow that he'd agree to accept me, and the deal was sealed. The only thing I asked of my assignments officer was that he not call me up at the last minute and tell me he couldn't make the home base assignment back to Bragg work. He assured me that with Shacknow's commitment it wouldn't be a problem. Or so he thought.

I LOOKED FORWARD to my next assignment in Korea, working for Edwards. Little did I know that during the course of this tour, I'd stumble on some information that would, if it were true, cause a paradigm shift for all of us Vietnam era veterans. And General Downing would finally extract his revenge for the honest PT test I had given him years before. Or so he would think.

Chapter 9

Korea

J3 Special Operations Command Korea (SOCK)

THE FOURTEEN HOUR flight wore on my nerves as usual. I'd spent so many hours flying around the world I swore, that when I retired, I'd never again ride in airplane. Dan Edwards and his Korean driver met me at the Incheon International Airport in Seoul, Korea. We drove straight to the garrison at Yongsan, which housed the Headquarters, U.S. Forces Korea (USFK). I signed into the unit; then Dan took me to my little apartment which I'd call home for the next twelve months. It had a bedroom, bathroom with a small living room, and kitchen attached. Compared to some of the places I'd lived overseas, this was high cotton.

Carolyn and Dan had me over for supper the next night, and we got all caught up on what had been going on at Fort Bragg and with Polly and me. It was great to be among friends in this most strange land.

SOCK's mission was to plan and conduct special operations in support of US and Korean forces. Edwards was part of the 8th Army's CJ3, which was a combined and joint command supporting military operations in South Korea. He also served as the US Deputy commander for Combined Unconventional Warfare Task Force (CUWTF). LTG Suh, a Korean three star, was the CUWTF commander.

As Edward's J3, I managed all the operational planning for Joint and Combined Special Operations. I had an office with Edwards in Yongsan and an office at the Korean Special Forces compound, which sat about fifteen miles and at least a one-hour drive away. Driving around Seoul makes driving around Washington, D. C. seem like a jaunt down a country road. About every other trip across Seoul, we'd see someone swipe off a side view mirror. No one ever stopped. They just shook their fist in the air, made an obscene gesture, and plowed on through the traffic.

I had working for me a joint staff consisting of Army, Navy, and Air Force Majors and a Korean-American NCO. We'd start out the day with about one hour in the Yongsan office; then we'd pile into our van, and our Korean driver would drive us across town to the ROK SF compound. Here we would spend the rest of the day working on the war plans and coordination with LTG Suh's staff. To assist me, Suh assigned a Korean Major to be our interpreter. MAJ Chung's English was as good and often better than mine. Frequently he'd use a word that I'd pretend to understand but later would look up. Don't know if that said more about Chung or about me. Probably both.

To make the trip across Seoul bearable and educational, I used my copy of *Elements of Style* by Strunk and White to conduct grammar quizzes. I think my guys and I learned more about punctuation and the use of the English language during those trips than we ever did in school.

In addition to the duties mentioned, the Commander of SOCK also had operational control of a beefed up ODA known as DET-K or Detachment Korea. DET-K consisted of twelve senior NCOs, all of whom spoke Korean and a Major as the commander. They provided liaison to CUWTF's seven Special Warfare Brigades. These units' primary tasks included airborne operations, reconnaissance, unconventional warfare, information collection in enemy territory, and carrying out special missions. I've known some hard soldiers in my time, and the ROK SF guys ranked among the top.

CUWTF also had a special all female battalion. These ladies were trained in all the arts of war and were tough as the guys. And they weren't hard on the eyes either. I was never privileged to know exactly what their role in our war plans was even though I was in charge of writing and updating the plans. I once heard Chung refer to their mission as "Lure Operations." I didn't push him for details. It wouldn't have done any good anyway.

We worked five-and-a-half days a week, taking off only Saturday afternoon and Sunday. I didn't mind that since I had nothing else to do. On the off days, I'd usually walk into Itaewon, Seoul's shopping district. The little shops that lined the streets bulged with colorful silk jackets, scarfs, and ties. The prices were unbelievable. Of course they had all the knock offs, but they had some really good stuff as well. Every trip there were guys on the street, offering to sell me a "genuine imitation Rolex." I kid you not. That's what they'd say. You could easily tell a fake Rolex from a real one. The fake's second hand ticked around the dial while the real one's second hand moved smoothly around the dial. I'd ship coats, scarfs, pocketbooks, and more back to Polly, Arlene, Sid Shacknow's wife, and to Joanne Lauder, Shacknow's secretary. Joanne was a small, thin lady. Her light brown hair framed a pretty face. She'd been around the command quite a while and knew where ALL the bodies were buried. Years later when she finally retired, there were more stars in the Officers' Club's banquet room than in heaven. She was quite a lady and one you definitely wanted on your side. She liked me. A little bribery would go a long way.

I was on jump status, which was great. The only problem was that we had to make our quarterly jumps from a balloon. The drop zone sat surrounded by sheer rocky cliffs on three sides that rose 1500 feet. The balloon was shaped like a small blimp. The metal gondola which hung under the balloon was open all the way around. It would creak and groan and sway as we floated up, tethered to a quarter-inch cable. On the way up to the 1000 foot level, we would hook our static lines into the static line cable, and check our equipment. It was deathly quiet. You could speak in a normal tone of voice and hear everything being said. Absolutely weird. Finally, the jump master would open the little gate with a bang, and we would just step out into nothing, falling straight down feet first.

This whole thing was extremely uncomfortable for me. By then I'd had over 200 military and civilian jumps. I didn't know how much I'd miss the security of being surrounded by the aircraft's skin and hearing and feeling the engine's roar. I even missed the blast of air that would push my feet up and sometimes over my head.

And of course there were all the small boulders waiting for me on the DZ. It's a wonder I didn't break my leg. If it hadn't been a macho thing, I think I would have asked to be taken off jump status. Hey, $110.00 a month didn't mean *that* much to me at the time.

While I was there, Polly came over for a two-week visit. LTG Suh's wife, Sookja, actually went to the University of Georgia at the same time Polly was there getting her Masters, although they didn't know one another. This small world thing instantly bonded the two. Dan and Carolyn also made the visit even more special for her.

The Koreans are big on partying, and any excuse would do to bring out the food and booze. Suh and Sookja threw a big party in honor of Polly's arrival and visit. It made her feel very special. Sookja even took Polly to a real tea room and gave her a tour of the Women's University where she taught.

ONE DAY WHILE I was sitting in my office out at the ROK SF compound, MAJ Bates, my Army plans officer, poked his head in my door. "Sir, got a minute?"

I motioned for him to come in and sit.

He was too excited to even consider sitting. Instead, he rushed over and handed me several sheets of paper stapled together. "You're not going to believe what I just found in the back of a drawer in my classified file cabinet. It's a briefing. No author noted. Not even a classification. If it's true, its gotta be Top Secret."

I began flipping through the papers. The first page simply read, RADE; the next, MISSION. The following page listed the OBJECTIVES: Hanoi City, Hoa Lo Prison, SonTay Prison, CAS Site #39 Laos, Haiphong Harbor. I continued flipping pages until I got to ASSETS.

Bates, standing beside me, stabbed the list of four named Americans with his finger. They were credited as to have:

–Confirmed information on prisoners
and
–Identified 9 out of 13 prisoner of war camp locations.

Stunned, I sat back in my chair. I recognized all four names, but one flew out at me like a witch on a broom. If I were to believe what I was reading, which I did not, and it were to get out, it would cause a paradigm shift for all of us Vietnam era veterans. The name printed boldly in front of me was Jane Fonda! Fonda? She was listed as a "Planted–Radical Activist." Along with Fonda was listed Tom Hayden. Unbelievable. Further listed only as "Radical Activist" were Peter and Cora Weiss.

I can only surmise that the author of the briefing had to be COL Bob Howard. Howard was the SOCK commander that Edwards replaced. He was a Medal of Honor winner from Vietnam, and he was also reportedly in on the planning and execution of the SonTay Raid. The details of the raid found in the briefing were too specific for it to have been written by anyone other than someone intimately familiar with the operation.

To this day, I still have a very hard time believing what I had read. But why would anyone write such a thing if it didn't have some basis of truth? I suspect that the four individuals were somehow duped into revealing the information cited. Also cited was a North Vietnamese whose name and position I will not say.

MY YEAR IN Korea was drawing to a close. Only three weeks to go. I was looking forward to getting back to Polly and Fort Bragg and working for Shacknow at SF Command. I was sitting in my little apartment after work one day looking out the window at the black and white magpies pecking in the front yard for their supper. The phone rang, and I picked it up. "Lieutenant Colonel Davis."

"Sir, MAJ Jackson." Jackson was the guy who had set up my assignment back to Bragg. "I've got some bad news."

I took a deep breath, dreading what was coming.

"Your assignment back to Bragg has fallen through."

I squeezed the phone so hard it's a wonder I didn't break it. "Jackson, the only thing I asked of you was not to call me at the last minute and tell me I wasn't going back to Bragg."

"I know, sir, but it looks like General Downing might have pulled the plug on it." Downing is the guy I'd given the honest PT test when I was a Major in 1st SOCOM's G1, and Downing was a BG and Deputy Commander of 1st SOCOM. Downing now sported the four stars of a general and was in charge of USSOCOM (the United States Special Operations Command). This was the command that all the services' Special Operation Forces fell under. It was highly irregular for him to be inserting himself in personnel actions of any of the different services. But you know what they say about pay backs. . ..

"So you're telling me that there're no slots on Bragg for me. Any at all?"

If I'd been told that I couldn't get back to Bragg six months ago, I could have put in my retirement paper work and found a job in the area. But not now.

"Sir, I've been looking. I do have an instructor's slot open at the War College." The Army War College was located in Carlisle, Pennsylvania. I'd finally finished the correspondence course. It would be a cushy job and only about six-and-a-half hours from Bragg. I could make it back and forth until I'd put in the two years required to retire as a full Colonel. "I've got to tell you things are tight at Bragg now. I was just talking to LTC Ballard at the 18th Airborne Corps' Officer Management and he—"

"What? Hold it. Did you say Ballard? That wouldn't be Rick Ballard would it?"

Jackson's train of thought temporarily halted, and he said nothing.

"Check and see if Ballard ever had a tour at 1st SOCOM."

"I know he did." Jackson shot back. "He was a Major at the time."

"Okay, you hang up, then call Ballard back. Tell him that Tom Davis needs a job on Bragg and to make it happen."

I explained to Jackson the connection I had to Ballard. He'd worked for me when I was the Deputy G1. Jackson said he'd get right on it.

Three unbearably long days later I got the call. Ballard had contacted the 2nd Army Headquarters in Atlanta, Georgia, and got them to accept me into the Readiness Group at Bragg as their Chief of Combat Arms. The slot called for a Lieutenant Colonel, but the commander, COL Metelko, would be retiring in about six months. If it worked out, I would follow Metelko as the commander. I was in a promotable status but hadn't pinned on my Colonel's rank yet.

"So, whatta ya think? Want me to lock it in?" Jackson said.

"Absolutely! I've always wanted to be assigned to a Readiness Group!"

I hung up the phone and collapsed back into my sofa. *Now, what all the hell do you reckon a Readiness Group does?* I thought.

Sookja even took Polly to a real tea room and gave her a tour of the Women's University where she taught.

The Koreans are big on partying, and any excuse would do to bring out the food and booze.

Chapter 10

Fort Bragg, North Carolina–Again, Again

Readiness Group Bragg

I FINALLY ARRIVED home after the long, long flight back from Korea. I drove out to Bragg and signed into RG (Readiness Group) Bragg and had an office call with COL Metelko. Metelko had been the 18th Airborne Corps G1 before taking over RG Bragg, so he was well tied into both Bragg and 2nd Army in Atlanta, Georgia.

RG Bragg's mission was simple–provide advice and assistance as requested to US Army Reserve and National Guard units in North Carolina. Our higher headquarters sat in Fort Gillem in Atlanta, Georgia. Since we were a tenant unit on Bragg, we didn't have to worry about any support tasking from the Corps or the installation. The 2nd Army Commander, a three-star general, and his one-star Deputy had to cover all the states in the southeast, so visits from either of them were rare indeed. I think during the entire time I served in the unit, we received only one visit from the commander and two from his deputy. We were our own little fiefdom. This had the makings of a great assignment.

The RG called home two two-story wooden World War II buildings and consisted of a headquarters and three divisions–Combat Arms, Combat Support, and Combat Service Support. A Lieutenant Colonel headed each division with Majors and senior NCOs, representing most of the various branches of the Army under them. This was necessary as we provided advice and assistance for Guard and Reserve units from all branches.

Our major customer was the 30th Armored Brigade Combat Team (ABCT) headquartered in Clinton, North Carolina. They had units that stretched from Manteo on the coast to Murphy in the mountains. The reserve command headquarters for North Carolina Reserve units sat on Fort Jackson, South Carolina. The North Carolina State Adjutant General, MG Rudicil, headquartered in Raleigh. To cover all these units, the RG soldiers were constantly on the go.

My division, Combat Arms, consisted of an Infantry Lieutenant Colonel, an Artillery Major, and an Armor Major. Each had a senior NCO of the same branch. I also had a civilian secretary who typed all reports and managed our TDY budget. Since most of us spent Saturday and Sunday visiting units, our weekend fell on Tuesday and Wednesday. We would spend Thursday and Friday preparing for and driving to the upcoming weekend's visits, returning late Sunday or mid day on Monday. We wrote our Trip Reports on Monday.

It proved a hectic schedule, but whoever was on Bragg and whatever day it was, we had a PT formation at 0700. I hadn't done PT as a unit in, well, forever. That took some getting used to. I worked out twice a day, so lunch time found me in the pool swimming laps.

I'd always held the soldiers of the Guard and Reserve in high regard, but it wasn't until I worked closely with them that I came to understand the dedication required of these fine soldiers. In the case of the officers and senior NCOs, there was no way that they could get all done that needed doing one weekend a month. Many, if not most, worked at their military jobs every week to some degree or another. The two weeks of AT (Annual Training), required several weeks prep prior to and after to ensure a successful operation. These weeks never got counted as duty time.

ON 1 FEBRUARY of 1992, Polly and COL Metelko pinned on my Colonel's Eagles. I moved from chief of Combat Arms to the Group's XO. LTC Drew Zegler, my Infantry chief, took over the Combat Arms Division. Metelko would be retiring in a few months and being XO made for a smooth transition.

WE SAID GOOD by to COL Metelko. I hated to see him go. I had found him easy to work for. His knowledge of the units we advised and his pull on Fort Bragg would be sorely missed. However, I had an ace in the hole. Mr. John Sheets, a retired Lieutenant Colonel I had known when he was the G4 of 1st SOCOM, was a senior civilian who worked in the Combat Support Section. He'd been with the RG for several years since retiring from active duty. He became not only a great friend but a trusted advisor. He could and would walk into my office and let me know exactly what he thought about anything that we had going on.

I gathered all together and went over the Five Pillars of my command philosophy. The RG proved to be a very different type of command than any I'd ever had. About half of all my soldiers were females, and all of the secretaries were female. The Army had long since made a big deal of sexual harassment. I'd never thought too much about it since I'd never had any females to harass. To be truthful, RG Bragg didn't really have a problem. But just to be sure, when I gathered all together I said, "No one should feel uncomfortable in the work place. Now I've got to tell you that hearing cuss words in mixed company makes me uncomfortable. That being the case, our new rule is that when the guys are together, they can say anything they want. When the gals are together, they can say anything they want. But when we are mixed, there will be no cuss words. I expect that we will all slip up every once and a while. When that happens, the person slipping up will immediately apologize to all in the room."

One of my female sergeants raised her hand. "Sir, exactly what are you considering cuss words? I mean like hell or damn? Everyone says that."

"I'm not going to make a list of what is acceptable and what isn't, so let's make it simple and forbid any and all cuss words. That way there will be no misunderstandings."

We had plenty of slip ups in our weekly staff meetings, but apologies were quick to come. I never had a single sexual harassment complaint during my entire time with the RG.

I HAD LONG ago realized that computers were the wave of the future. Currently, most in the RG hand wrote their trip reports and gave them to their secretary to type. Same with briefing, which the secretaries would put into PowerPoint slides. I procured three laptop computers with dial-up modems for each division and one for me. I had my IMO (Information Management Officer), a young SP4 female and my Signal Officer, a Captain, set up training classes in Windows (Microsoft had just moved from a DOS-based operating system to Windows 3.0), Word Processing, and PowerPoint. I then dictated that secretaries would no longer type trip reports or briefing slides. The RG's soldiers cried like rats eating onions, but their complaints fell on deaf ears. We even mastered sending briefings and files via FTP from one computer to the other over the phone lines. This doesn't sound like much now, but remember, this was before e-mail and the Internet, if you can remember back that far.

I SPENT MOST of my time visiting the TAG in Raleigh or the Reserve Command Headquarters at Fort Jackson in South Carolina. I got to know the TAG, MG Rudicil, well. The governor appoints the State AG. Sometimes it is an "Old Boy" deal, but not in the case of Rudicil. He'd been a platoon leader in Vietnam and had commanded at every level up to and including the 30th Brigade in the NC Guard. I found him to be an excellent commander. The troops knew and respected him to the max. This made my job not only easy but a joy to have.

Many was the time I thought about sending General Downing a note thanking him for screwing me out of my assignment back to SF. I even had the letter written. Polly read it and told me that I was being tacky. She was right as always. I wadded it up and tossed in the trash.

ONE OF THE nice things about having your own fiefdom was that you could do just about anything you wanted. You know, as how I was the King. This was the first unit I'd ever been in that I wasn't required to be ready to pack up and go to war with only a few hours or days notice. I had a couple of single parents working for me. In light of that, I decreed that single parents didn't have to come in for the 0700 PT formation. Instead they could do their PT during lunch. I wanted them to be able to get their kids off to school like normal parents would.

That announcement immediately bought forth a question from LTC Zegler, my Combat Arms Chief. "Sir, what about those of us who aren't single parents? Doesn't this violate your command philosophy about being impartial?"

He had a point. "Not really, as this rule applies to temporary single parents. You know, if your wife or husband has to go visit his or her parents for some reason and you're there by yourself with your kids. You can do your PT during lunch until he or she gets back."

All agreed that this was fair. And the King ruled again.

I DECIDED TO resurrect the two-day Gut Check event that I imposed on my officers and NCO of the 2nd SF Training Battalion. Remarkably, the guys and gals in the RG were more excited about participating than the SF guys were. Go figure. My old CGSC friend Monty Montero, now a General and in charge of 18th Airborne Corps' Corps Support Command, wanted to give it a try. He also brought along his aide decamp to participate in the fun. My brother John got wind of the happenings and wanted in as well.

I remember that I had a black female NCO who couldn't swim, but was damned and determined that she would take on the Gut Check. This presented a problem as one of the first day's event was the thirty-minute swim. Not to be deterred, she suggested that she "swim" in one of the lanes next to the side of the pool. After the push-ups, sit-ups, and sixteen minute run, she jumped in the pool and pulled herself down and back hand over hand for the thirty-minute swim. I don't think I've ever been more impressed with anyone before or since.

Montero broke the record of number of pushups completed in two minutes. My brother John came in second in distance swim in thirty minutes. I held the record there. Montero's aide blew by me in the last ten minutes of the ninety-minute run. He's sneaked up on me, and when he passed, I didn't have enough left to catch up. It's noteworthy to cite here that the next day when John took his first step with the forty-pound rucksack to begin the six-hour walk/run, it was the first time he'd ever had a rucksack on his back. And it was the last. All in all, good times were had by all.

WHILE I SERVED as RG Commander, a decree came down from the Army that active-duty forces could be used in support of States under a declared emergency like a hurricane or earthquake or such. Someone had the good sense to suggest that in that event, the RGs would assume OPCON (Operational Command) of those active-duty forces in support of the State's National Guard. This made perfect sense. The RGs were structured such that they had all the different branches represented. More importantly, the RGs had close working relationships with all the States' National Guard units. During a State emergency, the Guard could be called up to support rescue and recovery efforts. Active duty via control of the RGs would assist where needed.

This served to give a RG as close to a real military mission as could be hoped for without actually deploying into combat. I structured RG Bragg for this mission just as a Special Forces Group would for combat. I had an Operations Center, a Support Center, and a Signal Center. Each with the same functions as those of a SF Group when it went to war. When we were activated, I set up in the State's Emergency Operations Center in Raleigh. When a requirement came up that the Guard couldn't fill or would be better filled by active-duty personnel, I'd validate the request, then send it to my point of contact at Fort Bragg. If Bragg approved, the active-duty unit/personnel would report to me, and I'd further assign them to perform whatever task we had.

While I commanded the RG, we deployed in support of MG Rudici's National Guard for two hurricanes and one severe snow/ice storm. Because of the close working relationship we had with the National Guard folks, every operation went well.

I SAT in my office plowing through the latest trip reports when Cindy, my secretary, stepped into my office. "You've got a call. I think it's your dad. Line one."

I picked up the phone. "Dad, what's going on down there in wonderful Vienna?"

He didn't draw it out, rather got right to the point. "Johnny's had a heart attach and has died."

I reeled back in my chair, feeling the blood leave my head. Johnny had gotten out of the Army to help his mother and brothers run the Georgia Pacific Pump Company. Johnny loved the Army and had a bright future there. He'd been awarded a Bronze Star and a Purple Heart along with other awards. I thought then and still do today that if he'd stayed in the Army with its mandatory physical fitness requirements and its annual physical exam, we'd be sitting around drinking beer and reminiscing old times today.

Dad said that Joyce, Johnny's wife, wanted me to give the eulogy. I would do this. When the time came, I stood in Vienna's First Baptist Church and gave a difficult and heartfelt final farewell to my first best friend. Johnny has stepped in and out of my thoughts over these many years. I miss him with a pain that never goes away. Recently, I wrote a poem about our friendship and will share it with you now.

Best Friends

A fat summer moon
painted the ribbon of asphalt
with a liquid silver.

The tantalizing scent
of green peanuts, cotton dust,
and honeysuckle
drifted across the highway.

An early dew beaded our eyebrows.
As we peddled our bicycles into the night,
the broken white center line clicked by.

The Most Fun I Ever Had With My Clothes On

We were young then
and tasted the sweet freedom of youth.
Sure that the world was ours;
ours for the taking.

A blood oath we took.
Friends forever.
I and my best friend, Johnny.

Too soon,
taken away from me
by a sick heart.

I HAD BEEN back at Bragg and with the RG for two-and-a-half years when I got a call from Colonel's Assignment Section in MILPERCEN. I don't recall the assignments officers name, so I'll call him LTC Blockhead.

"Sir, we are looking at reassigning you shortly. Just wanted to call and discuss some options."

Colonel assignments were usually negotiated between the Colonel and his assignments officer. Generally speaking, a Colonel wouldn't be assigned to a command unless the first General Officer in the command agreed to accept him. Just like when Shacknow agreed to accept me as his Chief of Staff. It helped greatly if that General Officer called the Colonel Assignment Section and made a personal request. So far as I knew, no one had asked for me to be assigned to them.

"And just where were you thinking of sending me?" I was caught off guard and needed time to think. I swivelled around in my large brown high-backed executive chair and looked out the window.

"I have a position open at Langley that you might be interested in." Langley was the headquarters for the CIA. Sounded interesting, but it was near Washington, D. C.. That was a non-starter for me.

"How much time do I have to decide?" I had just about twenty-five years in the Army and a little over two as a Colonel. I really wanted to make it to thirty years as the retirement check would be substantially larger than if I punched out with only twenty-five years. If I waited until I received reassignment orders, then put in my retirement paperwork, I would have to get out in three months. They called this "retirement in lieu of orders." On the other hand, I could put my retirement paperwork in now and retire one year out. This would put me over twenty six years, the last pay increase before the final increase at the thirty year mandatory retirement date.

Blockhead said that he'd hold the Langley job for me for a couple of days.

I hung up the phone, frowning.

John Sheets poked his head in my door with a question, but stopped mid sentence when he saw my face. "Uh oh. Who's screwed up now?"

"Nobody. Come in. Have a seat. I need some advice."

Sheets walked in and sat on the big brown sofa to the left of my desk and crossed his legs. I explained the conversation I'd just had with Blockhead.

Sheets poked out his lower lip and frowned, then his face lit up. "Rudicil's Senior Army Adviser is on orders. I haven't heard of anyone taking his place." The Senior Army Adviser to the TAG was an active duty full Colonel. "Why don't you give Rudicil a call and see if he'll ask MILPERCEN to assign you there."

I nodded. It would mean that I'd have to get an apartment in Raleigh, but that was better than moving to D. C., and I could make it home most every weekend. "Good idea!"

I finally got through to Rudicil and explained my situation. He readily agreed to accept me and told me I could tell Blockhead that he would call and request me for the job.

The next day I called Blockhead and told him I had a job in Raleigh.

I could hear him shift in his chair. Then he said, "That's not going to work for us. You've been around Bragg since 1982. Time for you to leave."

This was putting it bluntly. "Excuse me. Two years ago I had a tour in Korea. Last time I checked that wasn't on Bragg."

"Well, that doesn't count."

Then it hit me. "What you're really saying is that the Army hasn't jerked my family around since 1982 and it's time for them to move."

I think I caught him off guard because he agreed that that was it. I hung up the phone and immediately put in my retirement paperwork with a separation date one year later on 1 April–April Fools Day. Two days later I called Blockhead back and told him I was retiring. I think that was his plan B anyway.

BILL TANGNEY HAD returned to Bragg as a two-star general and now sat in command of Special Forces Command. Polly and I had gotten together with Bill and Kathi just about every two weeks, so he was up on what was going on with me. I was still radioactive (that's the way Tangney described my status) so far as SF was concerned. Downing remained in command of USSOCOM, so no one could get me assigned back into Special Forces.

I called Tangney right after I'd told Polly about putting in my retirement. "Well, you've got a year. Lots can happen in a year." I wasn't sure what Tangney meant by that, but I could almost hear the wheels turning in his head.

ABOUT THREE MONTHS after putting in for retirement, Tangney called. "Wanna guess who's punching out?" Before I could guess, he said, "Downing!"

"Well, I can't say that I'll be sorry to see the son-of-a-bitch go."was the first thing in my mind and out of my mouth.

"Yeah, well, now I can move on getting you back over here." "Over here" was Special Forces Command.

"But I've got a retirement date of one April."

"Not a problem. I'll call MILPERCEN and tell them I've got to have you and that they should push your date back to 31 October. It'll be in the same FY (Fiscal Year), so they shouldn't have a problem."

Within three days, I got the call from Blockhead. He wasn't happy, but when a two-star general calls Colonel's Assignments Section and says he wants a guy even if it is only for nine months, they have to roll over.

I told Blockhead "thanks for nothing"; then hung up. He still thought he'd won even though my retirement date had been moved four months later. Little did I know at the time, but Tangney had a plan, and Blockhead really wasn't going to like it.

RG Bragg's mission was pretty simple–provide advice and assistance as requested to US Army Reserve and National Guard units in North Carolina. From right to left: Myself, 30th Brigade Commander, Brigade XO, Brigade CSM.

On 1 February of 1992 Polly and COL Metelko pinned on my Colonel's Eagles.

From left to right: Johnny, my first best friend, Barbara, my first girlfriend, Me, Jana, my first cousin. I miss Johnny with a pain that never goes away.

SLDC Special Forces Command

WITHIN TWO MONTHS, my replacement arrived, and I left the RG and signed into USASFC (United States Army Special Forces Command). When I first returned from Korea, SF Command filled the old Delta compound. Before Delta took over the building, it housed Fort Bragg's confinement facility, complete with jail cells. Now SF Command found itself on the second story of a brand new three-story building. In the floor above SF Command sat the three-star general who commanded USASOC (United States Army Special Operations Command). Assigned under USASOC were all Special Forces Groups, active duty and National Guard, the Ranger Regiment, Task Force 160 (the special operations aviation brigade), SWC (the Special Forces School which also provided instruction in Civil Affairs and Psychological Operations), the 96th Civil Affairs Company, and the 4th Psychological Operations Group.

The Army's Special Operations Forces proved their worth during the first Gulf War, and money and resources flowed their way. Chances of advancement for officers and noncommissioned officers did as well.

I reported in to my old and dear friend and mentor Major General Bill Tangney. He smiled and said, "Welcome back to the Force, Tom."

"Great to be back. And thanks, by the way, for making it happen. Even if it is only until October." I eased into one of the comfortable chairs that sat to the right of his aircraft carrier-size desk.

Tangney flashed one of his quick smiles. "A lot can change between now and October."

I didn't know for sure, but I thought I detected a hint of a plan. No way. I had a fixed retirement date and no one could unfix that. "So. I've got no desk, no computer, no permeant pass to get in and out of this monster building. What do you have planned for me? Can I kick back and just do PT three times a day and enjoy life?" I knew better, but I had to poke at him.

"Right. Joanne will show you where to set up. See the IMO about a laptop. What you'll be doing for the undetermined future is whatever I want you to do." Another quick smile. "Basically, I need a guy to conduct investigations both formal 15-6 and informal ones when allegations involving my Lieutenant Colonels and Colonels pop up. You might also have to attend some functions that I don't have time for." By "don't have time for" he meant that were dog-and-pony-show-the-command's face, and smile proudly. I made a mental note to ask Polly to get my dress greens cleaned.

"Sounds fine to me." Actually the only thing that sounded fine to me was the investigations; the rest I could live without.

We shot the bull a while longer then Joanne stuck her head in the door and reminded Bill he had someone waiting to see him.

I got up and walked out. "So, Joanne, the General says you're the one to fill me in on where I'll be hiding out around here."

"You got that right, Tom. Follow me." She marched me down the hall and around the corner to a little hole in the wall near the elevators.

I glanced around at the desk and two chairs and a small three-tiered book case. "Looks like home to me."

With active and National Guard SF units scattered from Washington State to Florida and around the world, someone was always getting into a fix. I investigated everything from a training fatality to the misappropriation of funds during an MTT (Mobil Training Team) in Africa. I could type about as fast as someone could talk, so I recorded all the interviews myself. Additionally, I went to meetings with Kiwanis clubs around the area and briefed them on what all we had going on. At least the open source stuff. I actually enjoyed these visits, although I didn't let on to that.

Shortly after I settled into my routine of PT twice a day and polishing up my investigative reports, I started signing the little yellow DA Form 61, buck slips, that I'd send into Tangney as Colonel Tom Davis, SLDC. I even hand printed a little sign and hung it on my door with the same.

I could tell it was driving Joanne crazy. As the General's secretary, she knew every acronym in the command and SLDC wasn't something she recognized, but loathed to ask me about it. Finally, she broke down. "Okay, Tom, I've gotta know. Just what the hell does SLDC stand for?"

"Well, it's a new title I created for myself, Shitty Little Detail Colonel." It broke her up. By the time I got back to my office, Tangney had sent for me. Within the hour the sign on my door came down, but I kept signing my notes as the SLDC.

OCTOBER AND MY retirement date loomed on the horizon. I hadn't done anything about another job for when I would get out. Every time the subject came up, Tangney would wave me off. Not to worry. So I hadn't.

Right after the first of May Joanne walked into my office and said, "The General needs to talk with you." She usually phoned me. A face-to-face invitation from Joanne couldn't be good, but her smile made me think maybe. . ..

I followed her down the hall to Tangney's office. "Okay, who got caught sleeping with who?"

"Nobody." Quick smile. "I've got an offer you can't refuse." He slid a paper across his desk to me that I recognized as a G3 tasking. I picked it up. It came from the Department of the Army through USASOC to USASFC requiring the command to cough up a replacement for the Colonel who presently served as the JSOTF (Joint Special Operations Task Force) Commander in support of OPC II (Operation Provide Comfort II). OPC II currently enforced the "No Fly" zone over northern Iraq. The tasking covered six months beginning in one month.

"Me?"

Tangney nodded and held his smile for a long two seconds.

"But this is for six months. I wouldn't get back until the first of December. I'll be a civilian starting on one November."

"Well, it's like this. The tasking is a must fill, and I've only got a week to come up with a name of a Special Forces Qualified Colonel with prior command experience. You're the only guy around who fits the bill. I've got an office call with LTG Schoomaker in thirty minutes."

LTG Peter Schoomaker currently commanded USASOC. He was an old Delta trooper and an outstanding soldier. I'd met him several times and really liked him, but asking him to go to DA and insist that my retirement be pulled was asking too much.

"You think he can get DA to pull my retirement?"

Another quick smile. "We'll see. After all, you're the most qualified Colonel to perform this most essential and demanding mission in a combat zone, which will require a mature and seasoned Special Forces officer."

"Well I hope when you talk to Schoomaker about it, you'll cut back on the sarcasm."

A quick smile. "Why, of course."

I looked at him. "Things like this don't just pop up out of the blue. You've known about this tasking a long time. It's a reoccurring one." I smiled. "You had this in mind when you pushed my retirement out to October."

An even quicker smile than normal. "Why, of course."

SCHOOMAKER BOUGHT IT and called DA. When a three-star Commanding General wants something, he gets it.

Not three days later my phone rang. It was my assignments officer, old Blockhead. "You think I'm going to let you pull this crap on me? You've got a retirement date that has already been moved once. I'm going to have to nonconcur with pulling it completely."

I couldn't help smiling into the phone. "Blockhead, you do what you gotta do, and I'll do what I gotta do." I then hung up on him.

ONE MONTH LATER, I left the Raleigh airport on a Delta flight, and I winged my way seventeen hours across the Atlantic to Istanbul where a staff car from the Incirlik Air Base picked me up and drove me to my home for the next six months.

Finagling my assignment to the JSOTF was the last thing Tangney did before moving up to command SWC, the Special Forces School. My old friend and fellow battalion commander in the 5th Group, MG Ken Bowra now sat as commander of SF Command. I used to tell him that one of the reasons he made General was because I was taking all the heat from COL Davis while we commanded battalions together. He and I had talked just before I left. He wanted me as his Chief of Staff when I returned and had greased this with Schoomaker. No more investigations for me. Now, I'd be appointing someone else to do them. Or so I thought.

JSOTF Commander, Incirlik AFB, Turkey

THE JSOTF'S MISSION was to provide a recovery/extraction force for military (members of the MCC– Military Coordination Center. More on that later) and "civilian," read that CIA, personnel who were deployed in Northern Iraq as part of Operation Provide Comfort II. OPC II enforced the no-fly zone over Northern Iraq via Air Force F-4 Phantom attack aircraft. This mission had been ongoing since the end of the first Gulf War. An Air Force one star commanded OPC II. It was a Joint and Combined operation incorporating US Army, Navy, and Air Force personnel as well as military representatives from Turkey, England, and France. Ultimate command and control fell under U.S. European Command (EUCOM), headquartered in Vaihingen, Germany. Special Operations Command Europe (SOCEUR), commanded by a one-star general headquartered in Stuttgart, served as my higher headquarters.

The JSOTF consisted of an Air Force Major serving as my J1, a Navy Lieutenant (equivalent to Army Captain) serving as my J2, an Army Captain serving as my J4, and an Army Special Forces Major serving as my J3. All had senior NCOs from the different Armed Services within their sections. We worked out of a large bunker near the air strip. The 10th SFGA supplied the J3 Major and a company of Special Forces soldiers commanded by another major. This company served as my quick reaction force, and rotated back to the 10th Group at Fort Carson, Colorado, every three months.

I quickly decided that going to war with the Air Force was the only way to go to war. I had my own apartment with a small kitchen, bathroom, and bedroom that also served as the living room. I even had a TV that brought in the Armed Forces Network. The 10th Group guys lived in air-conditioned tents. The Air Force provided a double tent that housed all kinds of weight and exercise equipment. They even supplied a large tent full of paperback books we could check out.

Incirlik Air Base sat a mere eight kilometers from Adana, Turkey. This was my second assignment to Incirlik. When I served with the 1st Battalion of the 10th Group in Bad Tolz, Germany, the American Red Cross sent me there to set up and run a three-week Aquatic School. I trained Air Force personnel from all over Europe as Water Safety Instructors.

Like most combat zones, days and weeks of waiting with little to do were interrupted by minutes or hours of crises. OPC II proved no exception. In fact, there was so little to do the company commander of the SF company asked if his guys could set up a mountain training course and offer it to any of the base personnel who wanted to take it. I readily agreed. The Air Force BG in charge of the base jumped on the opportunity.

I was ecstatic to find that the Air Force still maintained the Olympic size pool. I signed out a bicycle from Recreation Services and kept up my Triathlon training schedule with daily double workouts, combining biking, running, and swimming.

AFTER INTRODUCING MYSELF to the soldiers, sailors, and airmen of the JOSTF, my J3 and I set up a visit/recon to Zakhu (on some maps spelled Zakho), Iraq. From there we would visit Kani Bot, Iraq, a small Kurdish village located at the far eastern edge of the security zone in northern Iraq. We flew from Incirlik Air Base in a C-12 (small twin-engine fixed wing) to Diyarbakir, a transfer point in eastern Turkey.

Here we went through Turkish customs, then on to Zakhu, Iraq, by a UH-60 Black Hawk helicopter.

When we reached Zakhu, we landed inside the Command Post (CP), an area about the size of two football fields enclosed by an eight-foot high concrete wall. From there, we moved by yellow Toyota Land Cruisers outfitted with M60 machine guns and secured by armed Peshmurga guards to the Military Coordination Center (MCC) located at a place called the Zakhu house, also called Z-house.

The MCC's mission was to show a presence and monitor activities in the security zone, an area in northern Iraq that bordered Turkey well above the 36th parallel. The MCC was under co-command of a US Colonel and a Turkish Colonel. (Some of you older Special Forces guys may recall that COL Jerry Thompson, Special Forces, was killed by friendly air-to-air fire when he was in command of the MCC) There were also French and British officers and NCOs along with a split (six-man) Special Forces Team that was detached from the Special Forces Company I had OPCON of. This mixture of multi-national armed forces constituted a Combined Command. Each day the MCC folks visited different Kurdish villages throughout the security zone. The visits were conducted alternately by air or ground.

On a typical village visit, the US and Turkish co-commanders, along with the French and British Lieutenant Colonels, would meet with the Mukhtar (mayor) and the village elders. They would typically discuss village history, water source, medical coverage, schools, access to village, security, agriculture, Non-governmental Organizations (NGO) reconstruction, and other topics as they arose.

OUR CONVOY FROM the CP to the Z-house brought us down "gasoline alley," a four-lane paved road where the Turks dropped off large red sacks of potatoes and picked up Iraqi diesel fuel for the return trip to Turkey. These "food shipments" of potatoes were allowed under the United Nation (UN) sanctions (food for oil), but guess what? Nobody wanted potatoes, so red bags of potatoes lay stacked and rotting along the highway. The sole purpose for this delivery was to get the vehicles into Iraq under the UN's humanitarian umbrella. Then the Turks could then fill their trucks with cheap Iraqi fuel that they sold for a huge profit in Turkey. This was no secret as everyone involved with OPC II and those above that level knew what was going on. Typical UN operation.

We reached the entrance to the Z-house, zigzagged our way through several barriers (large concrete culverts filled with sand), and stopped in the parking area. We arrived too late to go on a "road trip," so we unpacked and received a tour of the Z-house.

The Z-house was a three-story building divided into three parts, each identical to the other. The construction was typical of the area: square with a flat roof. The walls and roof consisted of thick concrete. Blue and white tile covered the floors. Stairs accessed the three segments of the house and zigzagged to each level, opening onto the roof. The inside consisted of sleeping quarters, mess area, lounge areas, and operations and communication areas. Guards with M60 machine guns and RPG rockets stood watch 24/7 at strategic points on the roof.

To its rear, the Z-house had a covered veranda that housed a ping-pong table and lounge chairs. White metal ceiling fans continually chopped the air, keeping the flies down to a minimum. A small swimming pool (about 20' X 30') sat to the left of the veranda.

Because we missed that day's village ground trip, we decided to go into Zakhu. Since half the townspeople carry AK47s, the MCC provided us with an armed escort. The MCC security folks assigned us three Kurdish guards, a French NCO named Terry who spoke fluent Arabic and English, and a Special Forces Captain, all armed with AKs. The people here, all Kurds, were extremely friendly toward us, but the MCC took no chances. Kids there were like kids anywhere. They loved to have their pictures taken, especially with US military soldiers.

It was a religious holiday, so the streets were not as crowded as usual. What can I say about Zakhu? I suppose it, too, was typical of the region. Small shops lined the streets, selling everything from gold to vegetables to cigarettes to rugs. Down the middle of each alleyway ran a stream of sewage. The market street was crammed with all manner of vegetables and meats and was abuzz with flies and with people trying to sell you something. Barter was the name of the game, and we let Terry negotiate for us. I bought a few post cards and an Iraqi bayonet, which I would have to sneak back through Turkish customs. We visited the older part of the city and saw an old Roman bridge built well before the time of Christ. The bridge crossed the river that ran through Zakhu and into the Tigris.

That evening I received a briefing on the operations of the MCC, and we had dinner. Assigned to the MCC was a British Army "chef" (he was quite adamant about the term "chef" as opposed to "cook."). After I ate one of his meals, I knew why.

❖

THE NEXT DAY after lunch, we received a briefing on our ground movement from the Z-house to the Command Center where two Black Hawk choppers would pick us up for our air mission.

We left the Z-house and a few minutes later arrived at the Command Center. From there we proceeded on a 45-minute flight to the far eastern end of the security zone. Northern Iraq looked exactly like much of our American West. We lifted off and, once clear of Zakhu, dropped to about 60 feet and flew nap-of-the-earth, racing toward the mountains. Sitting in the chopper with the doors open and sandwiched between two M60 door gunners, I flashed back to Vietnam.

We soon reached the mountains. Large outcroppings of a rock–similar to granite but a much lighter color–dotted the ground. We wove and bobbed our way through mountain valleys, following streams and dirt paths the Kurds call roads. The number of trees that covered the mountains slopes amazed me. All grew round and no higher than a basketball goal. As we snaked our way higher into the mountains, the temperature dropped, making me uncomfortably cold. Finally, the village of Kani Bot appeared in front of us.

I flew in the lead chopper. It circled to find a landing zone that was flat enough and large enough to accommodate it and the other chopper that accompanied us. On our first landing attempt, the chopper slid backwards as the rear wheel touched down. The pilot immediately lifted off. Our second attempt was no better. Finally, on the third try, we landed, immediately followed by the second bird.

We presented quite a show as we disembarked from the choppers. Surrounded by our Peshmurga guards, we moved toward the village. All the children, dressed in colorful clothing, followed us, smiling, waving, and shaking our hands. The villagers seemed, and in fact were, glad to see us.

THE MUKHTAR, SALIM Saleh, was not there, so we met with his number-two man and all the village elders. Kani Bot consisted of about ninety families. No one could give us an exact or even an approximate number of how many people lived there. The Iraqi army had destroyed this village five times (Saddam was intent on "cleansing" his country of its Kurds), the last time being in 1988. All villagers had fled to Iran, where they lived in a camp called Zewa. About half of them returned to Kani Bot in 1991. Sixty or so houses now spotted the area, all self-built, mostly of cut-stone with mud-based mortar. Large poles supported a mud roof. After a rain, the owner used a large stone–shaped like a role of paper towels but about three times its size—to roll across the roof. This procedure squeezed out the water and further packed the mud. The houses' floors were also packed mud. Since there was not enough housing to go around, families shared.

The village got its water from three springs. Unfortunately, they had no water distribution system, so the villagers had to go to the different springs and draw the water there. Their drinking water was separate from the washing area. The water supply was insufficient, especially during the summer. The quality of the spring water was good. We all took tea with the village elders during our visit. Not to do so would have been a supreme insult. I would pay dearly for this tea after I got back to Incirlik. Once again, I'd have severe diarrhea in yet another far corner of the world.

The village had only minimum medical support. In the event of a major medical problem, the patient would be carried either on foot or by mule to the nearest hospital miles away. Their major medical concern was malaria, especially in the summer when up to 50% were affected.

They had one teacher who taught up to grade three. After that, the children either went to school at another village or ended their education.

Farming presented a challenge since the village sat high in the mountains. Because of the natural slopes, lack of oxygen, and few mules, they could only farm a small part of the area. They mainly grew wheat, tomatoes, and okra. They ate most of what they grew. For the life of me, I could not see how these people survived the harsh winters and blistering summers, but they had for several thousand years.

When we entered the village, the Mukhtar's representative escorted us to a house. We removed our boots and entered a small room. Worn but colorful carpets blanketed the floor and walls. Here, with about fifteen of the village elders sitting with their backs to the wall, we discussed the state of the village. Children peered into the room through the bars that covered the larger of two windows. They served our tea in glasses about the size of a double shot glass. The tea tasted very sweet and very good. Everywhere you visited around Iraq or Turkey, the people offered you tea. They served it in grand style, presenting it in a little glass that sat on a small colorful saucer. A tiny spoon rested in the glass. You'd vigorously stir the liquid to get the half inch of sugar sitting on the bottom mixed in with the tea. I became quite fond of these occasions. Of course, the villagers offered to feed us supper, but we were pressed for time and couldn't accept. How they could offer us something that they had so little of amazed me.

We concluded our visit in about an hour and had our pictures taken with the village elders. We provided them Polaroid photos before we left.

I have rarely met a more generous group of people than the Kurds. I have also rarely met a more warlike race either. They have been fighting between themselves and with the surrounding countries for thousands of years, and I suspect this will continue for thousands of more years. I will say that Operation Provide Comfort II provided a stability to this region rarely experienced. I can't help but think that our presence in the area between the wars led to the success Special Forces experienced in Northern Iraq when the second Gulf War started.

I, THE SAME as you, heard cries from those who said that Bush lied, referring to WMD (Weapons of Mass Destruction). It's true we didn't find the large stockpiles of chemical and biological weaponry that Secretary of State Colin Powell assured us Saddam had. But let there be no doubt that the combined intelligence of the United States, France, Brittan, and Turkey supported this proposition. Those collective intelligence services also spoke of Saddam's desire and efforts to procure a nuclear capability as well. They were careful to assess that he was a long way from a nuclear capability though.

During my time commanding the JSOTF in Turkey between the two Iraq wars, I sometimes sat in what amounted to a Combined SCIF (Sensitive Compartmented Information Facility) at Incirlik Air Base and heard combined intelligence briefings during which representatives of the countries involved voiced the belief that Saddam possessed and was willing to use WMD to further maintain power and eventually dominate the region.

It's an absolute fact that Saddam used his chemical weapons on the Kurds living in northern Iraq. Given what I observed at the time, it is easy to see how our country's leadership concluded that Saddam was a threat that must be dealt with or the world might face a nuclear armed Iraq or at the very least an Iraq backed up by weapons that had long since been banned by the Geneva Convention.

Having said all this, you can see why I get irritated with those who claim to be an authority as to the veracity, or lack thereof, of President Bush concerning this issue. There is no doubt in my military mind that

President Bush and others in his administration spoke to what multi-countries' intelligence services believed true at the time.

So there.

LIFE WENT ON in the JSOTF. The Turks made it as hard as possible when the SF Companies rotated every three months. Just passing the Company's equipment through customs took four to five days. The Turks also restricted, to a great degree, what training the Company could conduct. It seemed like they were far more the problem than the solution.

I enjoyed the local people and eating in restaurants just off base. My favorite one had tables set on its flat roof. In the evenings, I could look out over the city as the Adhan "call to prayers" blasted from loudspeakers housed in the various mosques. Thankfully, the Turks had no problem with alcohol, so I enjoyed the local beer with hot cheese bread served before the meal.

I HAD BEEN in touch several times with MG Ken Bowra who currently commanded the SF Command. We talked about what I'd be doing as his Chief of Staff. He had both my room and office numbers. Around 1800 one day, my phone rang. I cut my TV off and answered, "COL Davis."

"Tom. Ken. Got some good news and some bad news."

Well, what now, I thought. "Give me the bad first."

"Our plans for you taking over as the Chief of Staff have fallen through."

Downing's name popped into my mind. "Don't tell me Downing's been called to active duty and is screwing with me again."

I could hear Ken laughing; then he said, "One, two, two, two, three, four, four, four." He was referring to the time when I'd given General Downing an honest PT test and would repeat the number of unacceptable push-ups or sit-ups until he got it correct then would call the next number. Bowra did this every time he saw me. It was his very favorite "Davis story."

"Okay, Ken, cut the crap and tell me what happened." I couldn't imagine how this had gotten screwed up.

"The CG has decided he wants you to do something else." The CG he referred to was LTG Schoomaker, Commander of USASOC. "That's also the good news."

"So?"

"He's decided you'll be his IG." The USASOC Inspector General job called for a full Colonel. It was a high profile job and one that would have to be approved by the DA IG. "I've gotta tell you that I think Tangney has something to do with this."

When I heard that, I knew there was no use in trying to get Schoomaker to change his mind. What I couldn't figure out was why would Tangney maneuver me into the IG job. I'd have to wait until I got home and ask him.

MY SIX MONTHS of fun in the sun and twice a day PT sessions had come to an end. I boarded a Delta flight back to the States and to Bragg. My second order of business would be to pin down Tangney as to why the USASOC IG job.

We worked out of a large bunker near the air strip.

The people in Zakhu, Iraq, all Kurds, were extremely friendly toward us, but the MCC took no chances. Kids there were like kids anywhere. They loved to have their picture taken, especially with US military soldiers.

USASOC Inspector General

I TOOK A week's leave when I returned to Bragg. During that time, I visited Tangney at his office in the Special Warfare Center and School, SWC, and asked why he wanted me in the USASOC IG job and not SF Command's Chief of Staff. He reasoned that with a three star asking for me and a three star (the DA IG) approving the assignment, it would lock me in for three years at Bragg. By then, I'd be too close to my thirty years for DA to move me. Sounded right, but I felt that he wasn't giving me the full story. And he wasn't.

MY LEAVE BEING up, I reported to the IG's offices housed in the old 7th SF Group's headquarters. It sat only a block up Ardennes Street from Tangney's office in SWC. The first guy I met was Mr. Carl Cook. Mr. Cook was my civilian deputy. He'd been in the IG shop since he retired as a senior NCO. He'd spent his career serving in ever-increasing admin positions within Special Forces. A stocky man, he stood a few inches shorter than my seventy-three. White hair topped his head. He always wore a white long-sleeve shirt and tie. He'd prove an invaluable advisor and friend over the next almost four years.

As the Inspector General for a major Army command, our main job was investigating charges made against various levels of command. We also scheduled and conducted IG Inspections. To help me do this, I had a military deputy, MAJ Roger Bartlett, several senior NCOs, and a goodly number of civilians. Most of the civilians had prior service. All proved to be very professional.

I also briefed all newly selected Special Forces Battalion and Group Commanders when they attended the SF portion of their Pre-command Course hosted by SWC. All newly selected commanders attended this course. Briefings by all the General and Special Staffs took up several days of the Bragg phase. The gist of my class consisted of the five principles of good management. I told all who would listen that most, if not all, of the problems commanders found themselves in could be traced back to a violation of one or more of these principles. Some listened with noticeable interest, and some didn't. I'd most likely be seeing those who didn't in my official capacity as the IG before they finished their time in command.

BECAUSE OF THE sensitive nature of what we did in the IG's office, I won't go into many details. However, there was one interesting investigation I did that took me from Bragg to Stuttgart, Germany, and on into Bosnia.

Tangney, now a three-star general and the USASOC Commander, sent me to investigate the discharge of a 9mm pistol carried by an SF Medic. He had shot a Navy SEAL in their Teamhouse. Go figure.

At the time Special Forces had teams scattered across Bosnia, serving as eyes and ears for the Combined Command which fell under NATO. The SF Teams were augmented with two to three SEALs each. They actually lived in the villages in rented houses and mingled regularly with the people there as well as the mayors and other village officials. Each Teamhouse had an armory along with a communications center. The Teams provided valuable and timely information to the Joint Special Operations Task Force, who would then pass it on to the Combined Command's Headquarters.

When the word first flowed up the chain of command about the shooting, it was thought that the men involved were horsing around playing "quick draw" and the SF guy shot the SEAL. If true, this would prove embarrassing to say the least. I was sent in to find out what had really happened and report my findings to the commander of Special Operations Command Europe or SOCEUR, BG Jeff Lambert. Jeff and I had commanded battalions at the same time but in different Groups. I had an in-briefing with Jeff in Stuttgart, got my in country ID made, then caught a C-130 out of Ramstein Air Force Base to Sarajevo, where the Combined Command and the JOSTF were located. From there I'd visit the Team at their Teamhouse and interview the SF Medic who shot the SEAL. I'd also interview the SEAL who was currently in a US medical facility in country.

When I arrived in Sarajevo, I checked into the Holiday Inn. The once grand building now sat scarred by small arms fire and shrapnel. I rode a creaking, shaking, and halting elevator up to my room on the fifth floor. After I stepped off the elevator, I decided that I'd take the stairs from then on. The room was small and devoid of color. No pictures on the wall or curtains on the window. Only a small bed, desk, and night stand with a bedside light filled the room. It did have a bathroom though. I'd stayed in a lot worse.

I first interviewed the SEAL. He was typical of the SEALs I'd known. Happy-go-lucky in a professional way. He told the story that the two of them were dry firing at people on the Teamhouse's TV, when the SF Medic, thinking his 9mm unloaded, had an accidental discharge (AD). The result put a hole in the SEAL's shoulder. He laughed as he described the look on his buddy's face when the pistol went off. Oh, those SEALs.

I WOULD BE in Bosnia for a couple of weeks during which time I visited several of the Teamhouses throughout the country, including the one where the AD had occurred.

On one day, we drove north through mountains shrouded in fog and snow to Bugojna, a small village that sat about 100 kilometers northwest of the once great and majestic city of Sarajevo.

Snow blanketed the ground as our two-vehicle convoy eased out of Sarajevo. As we rolled through the countryside, I flashbacked to scenes from a former tour with the 10th Special Forces Group in Bad Tolz, Germany. Had it not been for the festering wounds of war, I'd have thought we were cruising through Bavaria's rugged Alps and quaint villages.

The carnage created by the warring factions robbed the Bosnians of more than just their homes and land. Rarely did anyone we passed standing beside the road smile at us. The Bosnians seemed totally neutral to NATO's presence there. A sad quiet veiled their faces, reflecting the years of senseless conflict that had ravaged this once proud and pristine country.

We drove along treacherous icy roads in our armed convoy. The lead Land Rover, driven by the most experienced driver—on this day a Navy SEAL—would pass a car then radio back over a hand-held brick radio when it was safe for the trail vehicle, in which I sat, to pass. This system worked quite well. Even so, we rarely exceeded 20 KPH (12 miles per hour) for most of the five-hour one-way trip. While driving through the mountains, I'd look to my right and down hundreds of feet into the valley below. I remembered the NATO soldiers who died when their automobile slid off the road and down the mountain about this same time of year.

All along the narrow two-lane road that snaked through the mountains we saw evidence of ethnic cleansing—houses totally destroyed, not just pockmarked by gun fire. In the cleansing process, the opposing sides would place large mines inside each corner of a home and tie them in with detonation cord. When they set off the explosives, the charges effectively blew out the four corners, causing the house to cave in on itself. We would see rows and rows of houses destroyed, then rows and rows of untouched homes. Who invaded whom (Muslim or

Serb) determined which houses were left untouched and which were destroyed.

Even amidst this senseless destruction, I saw children pulling sleds, throwing snowballs, and chasing each other through the snow. The resilience of these youths amazed me.

I had also heard that this country had some of the most beautiful scenery in Europe. I saw evidence of this as we slowly made our way through the mountains. Ice-choked waterfalls tumbled down steep cliffs. Crystal waters raced over stream beds dotted with rock islands. Evergreens wrapped in snow produced a blanket of green and white that covered the valley floor. Shear granite-grey rock walls stood imposingly in the distance, a challenge to any mechanized force that might oppose them.

As we climbed higher and higher into the mountains, we discovered that a snow plow had pushed aside the snow and piled it high on the shoulders, further narrowing the passage. Passing trucks barreled by, edging ever closer to our Land Rovers. Often, forty-passenger Mercedes buses blew by inches to our left. The SEAL driving handled it as though we were taking a leisurely Sunday drive. I, on the other hand, held my breath, waiting to hear the sound of metal grinding followed by the eventual tumble, end over end, down into a ravine.

Each day during our travels, we saw wrecked vehicles lying beside the road, wheels up, like giant road-killed armadillos. The driver and passengers would invariably be standing around with their hands shoved in their pockets, shoulders hunched against the driving wind and snow, wondering what to do.

As we passed through the various towns, I was puzzled by laundry hanging above the balconies. How could anyone expect it to dry in the snow and rain and sleet and fog? Another curiosity was the number of people walking beside the road, day and night. I don't mean just in the cities but along mountain roads, weaving through miles and miles of rural nothing. Where did these people dressed in their dark clothes and wearing their sad faces come from? Where were they going?

Bosnia was a country with great contrasts. Why it had done these unspeakable things to itself is buried in centuries of distrust and hatred spewed out over the years by one ethnic group toward the other, all done in the name of religion. It is truly one of the most tragic things I have seen in my travels, even rivaling the sad plight of the Northern Iraqi Kurds. This carnage stood as a true testimony to the ignorance of mankind. However, before one turns up his nose in disgust, let's remember what happened in America between 1861 and 1865.

At the time, I thought what grand potential Bosnia would have if only the people could live together in some semblance of harmony. Only time would tell.

WHEN I INTERVIEWED the SF medic who had the AD, he parroted a story similar enough to the SEAL's to be believable. I could see he was still very upset with what had happened. I would write up the incident as an accidental discharge, while serious, not something that should ruin the guy's career. I believe he went on to later attend Physician's Assistant school.

Before I returned to Stuttgart to out brief the SOCEUR Commander, I dropped back by to check on the SEAL.

"So, how you doing?" I walked in and took a seat beside his bed.

"Great, sir. But they're sending me back States side." His face reflected his disappointment.

"That's okay. You'll be back with the SEAL Team. Plenty of crap to get into there. Just checking to see if you needed anything."

He waved me off. "So what's going on with the investigation?"

I told him what I could and that he didn't have to worry that his buddy would be in any big trouble. We talked a while longer then I stood up to go. Curious, I said, "So. What have you learned as a result of all this?"

The SEAL perked up, smiled, then said, "Sir, you know I've got me a 9mm back home for personal protection. As soon as I get back there, I'm selling it and getting me a 45. This little 9 didn't even knock me down. I gotta have me something with more fire power."

Those SEALs. You gotta love 'em.

I'D BEEN THE IG for about two years when Tangney came out on the list for his third star and replaced LTG Schoomaker as the USASOC Commander. At that time it was clear why he wanted me in the IG Job. I suspect that he knew, or had a pretty good idea, that he'd get promoted and replace Schoomaker. Now there really was no chance that DA would try and move me. I'd be at Bragg all the way to thirty.

I had an in briefing with the SOCEUR commander in Stuttgart then got my in country ID. On the back of the ID was written: The bearer of this card is a member of the **Peace Implementation Force**. All Civilian and military personnel are requested to extend him/her free access to the national territory and afford him/her all privileges in the execution of his/her duties. This text was in English as well as two other languages.

Retirement Times Two

ALL THE RUNNING, rucksacking, Triathlons, Ultra Endurance Events, and over 200 military and civilian parachute jumps had taken its toll. For my last year, I had a permeant P3 profile due to problems with my ankles, knees, and back. This was good news and bad news. The bad news was that I could no longer take the regular PT test with my soldiers, and I'd been restricted to water jumps only. The good news was that the alternate PT test I could take was the 800-yard swim. I had to complete it in 24 minutes. Most of my water jumps were onto Shark DZ down in Key West. Oh, throw Brer Rabbit in that briar patch!

Several months before I retired, I flew down to Key West and made a water jump. The next day I joined a five-person team consisting of MAJ Drake, the UWO Company's commander, his Company Sergeant Major, and two instructors. We competed as a team in the Swim Around Key West race. Each of us swam two one-mile links. We didn't come close to winning or even placing in the team competition, but we all had a great time making the effort. And I got credit for completing the Alternate PT Test.

MY RETIREMENT LOOMED only one month away. Mr. Carl Cook, my civilian deputy, had been pushing me to schedule a retirement ceremony on Bragg. I'd resisted. I just didn't want to have any fanfare when I left. The past thirty years meant so much to me that I was sure I couldn't make it through a speech without embarrassing myself.

I had my retirement orders in hand with a 1 August 1999 date in the upper right hand corner. Along with that the Army's Certificate of Retirement and a Certificate of Appreciation auto-pinned by the current President and my Commander-in-Chief, Bill Clinton.

I was still pissed with the conduct Clinton had displayed while in the oval office with a young intern. If a squad leader had pulled a similar stunt, he'd find himself kicked out of the Army with an Other than Honorable Discharge.

I suspected that I'd really like to sit down and have a beer with Clinton and maybe hit a couple of bars. I knew he was a very likeable guy. My old mentor and friend MG Sid Shacknow had met Clinton when Sid was a commander in Berlin. He was impressed with the guy. I just couldn't get past the intern thing. Oh well. . . .

The last time Cook mentioned a retirement ceremony, an idea burst into my head. Why not have my ceremony after a jump into Key West?

After all, this was a place that meant a great deal to me ever since I realized the childhood dream of becoming a Frogman there. So that's what I did.

I CALLED COL Andy Anderson, my old friend from my 5th Group days. He currently commanded the Training Group. I told him what I had in mind. He jumped on board, and told me that he had a C-130 headed down to Key West on Wednesday. Almost every month the Training Group scheduled a C-130 to fly to Key West so the instructors could get in their quarterly mandatory jumps. Today was Monday. He said he'd fly down with me and attend the ceremony. My next call was to the Company in Key West. I spoke with the Sergeant Major and asked him if he'd read my retirement orders after I swam in from the jump. He readily agreed. It was all set.

Wednesday arrived and we took off from Pope AFB and headed south. The weather was great. On board were a couple of other jumpers along with Anderson. We first landed at the Naval Air Station and took on members of the Company that needed a jump, and got the MACO briefing from the Marshaling Area Control Officer (MACO) then the jumpmaster inspection. After all, this was the justification for laying on the C-130s. Anderson had a U-21 scheduled to take him and me back to Bragg in two days.

When we touched down, I passed my retirement paperwork to the Company's Sergeant Major. He'd be waiting for me at the Dive Flag at the end of the swim. The jump went well. Floating down, I looked out over the DZ. The Under Water Operations School lay just off to my right and Key West to my far left. Above the horizon a tangerine sun beamed through spotty clouds. I thought back to when I'd been a student there and earned my Dive Badge, and declared myself a Frogman.

I hit the water and struggled out of the parachute harness. The safety boat recovered the canopy and harness, and I began the 1500-meter swim to the large Dive Flag posted just down the Key from the school.

When I got to the shore, the Sergeant Major stood waiting for me along with a few of the instructors who weren't on the jump or on the platform teaching. The Sergeant Major read the orders and handed me the Retirement Certificate along with the President's letter. I thanked him, then folded the letter into a paper airplane and sailed it out into the bay. I'm sure if Clinton knew what I'd done he'd lose a lot of sleep.

That evening I bought the beer at Sloppy Joe's. No speeches, no ceremony, no regrets.

THE FOLLOWING DAY, Anderson and I boarded the U-21 for our flight back to Bragg. In the past thirty years, I'd flown thousands of hours and jumped more than 200 times from various aircraft. I'd experienced two emergency landings. I figured that my luck flying had to run out soon.

When the U-21's wheels lightly kissed the tarmac at Hunter Army Air Field which sat next to Bragg, then began a steady roll down the runway, I leaned over to Anderson and said, "Andy, with any luck this will be the last time I ride in an airplane." It wouldn't be, but since that day, I've managed to avoid all flights except for two. I'd do my best not to repeat the experience. We'll see.

THE MONDAY AFTER our return on Friday from Key West, I got a call from Ruth Bellamy, the General's Secretary, telling me that LTG Tangney wanted to see me ASAP. I drove to the USASOC Headquarters, swiped my card to get into the building, then rode the elevator up to the third floor.

Ruth thumbed me in to Tangney's office. He sat behind his desk, his reading glasses perched on the tip of his nose. He glanced up at me and pointed to a chair. "Well, Tom, whatta we going to do about your retirement ceremony." He pointed down in the direction of the USASOC auditorium, which sat on the first floor.

I shifted uncomfortably in the chair. "Well, nothing. I've already had my ceremony. You. . . know. . . down in Key West. . . last week.

Tangney removed his glasses, sat them to his right, and flashed one of his quick smiles. "Not good enough. You're going to have a real ceremony and have it here. It's a way of bringing closure to a career. It's a good thing. It's the right thing."

I'd seen the look on his face many times before. It told me that I shouldn't fool around, just do what he wanted me to do, and get it over. Then an idea struck like a bolt of lighting. "Okay, I'll have a retirement ceremony, but not here in the USASOC Headquarters. I want to have it in the SWC auditorium. That's where I entered Special Forces, and that's where I want to leave."

Tangney liked the idea. I got up and started for the door. If I was going to put something together, I'd have to start immediately. I had only three weeks left.

As I approached the door, he cleared his throat. "And by the way. Want you to bring the President's letter with you. I'll read it to the assembled masses." He placed his glasses back on and gave me one of his quick smiles.

"Sure thing," I said. Tangney had spies everywhere.

As I walked past Ruth's desk, she smiled then said, "Can't wait to hear the General read that letter the President sent you." Apparently, Ruth had spies everywhere, too.

THE BIG DAY finally arrived and Polly and I sat on the SWC stage while Tangney stood at the podium. As usual, he gave a great speech which made me feel good. The best part was that he meant most of it.

When my time came, I walked up to the podium. The small auditorium was a little over half filled. I'd thought about what I would say. I wanted it to be something different, and I didn't want to choke up while I gave the speech.

So after I thanked all for attending the ceremony, I began, "I'm unique within Special Forces." Well, this got everyone's attention. But I figured it wasn't a brag if you could prove it.

I went on, "I have worn a Green Beret at every rank from Second Lieutenant through Colonel. I've commanded four A Teams, a Headquarters Company, two Battalions, and a Joint Special Operations Task Force. While in SF, I've seen duty on four continents and in ten foreign countries. I've been in and out of three combat zones. But that's not what makes me unique within the Force. What makes me unique is that having been through all that, I'm still married to the same woman."

My long march from Private to Colonel at an end, I extended my hand to Polly. She walked toward me, and I met her half way. When we met, I turned back to the audience and said, "If ever anyone earned the right to wear a Green Beret, it's Polly." And with that, I placed my beret on her head.

Everyone clapped as Polly and I marched down from the stage, up the aisle, through the auditorium doors, and out of the Army.

I hit the water and struggled out of the parachute harness. The safety boat recovered the canopy and harness, and I began the 1500-meter swim to the large Dive Flag posted just down the Key from the school.

My long march from Private to Colonel at an end, I extended my hand to Polly. She walked toward me and I met her half way. When we met, I turned back to the audience and said, "If ever anyone earned the right to wear a Green Beret, it is Polly." And with that, I placed my beret on her head.

~END~

Acronyms

Abn Airborne.
AD Accidental Discharge.
ADA Army Air Defense.
AFB Air Force Base.
AIT Advanced Individual Training.
AMMO Ammunition.
AOB Advanced Operations Base, a Special Forces tactical headquarters normally company size.
ARC American Red Cross.
Article 15 Non-judicial punishment given by a company or field grade officer.
AST Area Specialist Team, assisted an Special Forces Team while in isolation prior to a mission.
AT Annual Training.
BG Brigadier general or 07.
CA/PO Civil Affairs/Physical Operations Officer.
CAO Casualty Assistance Officer.
CCN Command and Control North, one of three command and control units SF had in Vietnam.
CGSC The Command and General Staff College.
CIB Combat Infantryman's Badge.
CIDG Civilian Irregular Defense Force, the fighters that SF trained and equipped to assist in conducting interdiction operations along the various borders of South Vietnam.
CJ1 The Armed Services designation for a personnel officer or section for a command consisting of members from different armed services like Air Force, Navy, Army and/or Marines and members of other countries.
CJ2 The Armed Services designation for an intelligence officer or section for a command consisting of members from different armed services like Air Force, Navy, Army and/or Marines and members of other countries.
CJ3 The Armed Services designation for an operations/plans officer or section for a command consisting of members from different armed services like Air Force, Navy, Army and/or Marines and members of other countries.
CJ4 The Armed Services designation for a supply/logistics officer or section for a command consisting of members from different armed services like Air Force, Navy, Army and/or Marines and members of other countries.
COL Colonel or 06.
CP Command Post.
CPT Captain or 03.
CSC Combat Support Company.
CSM Command Sergeant Major E9.
CUWTF Combined Unconventional Warfare Task Force.
DA Direct Action one of the Special Forces Missions.
DCO Deputy Commanding Officer.
DEROS Date Estimated to Return from Overseas.
DET-K Detachment Korea.
DZ Drop zone.
DZSO Drop Zone Safety Officer.
E5 Specialist designation for a sergeant, a buck sergeant. Also indicates a pay scale E1 through E9.
E6 Specialist designation for a staff sergeant. Also indicates a pay scale E1 through E9.
E7 Specialist designation for a sergeant first-class. Also indicates a pay scale E1 through E9.
E8 Specialist designation for a master sergeant. Also indicates a pay scale E1 through E9.
E9 Specialist designation for a sergeant major or command sergeant major. Also indicates a pay scale E1 through E9.

EOC Emergency Operations Center.
ETS Estimated Time of Separation.
FANK Forces Armee National Khmer, Cambodian Military forces.
FID Foreign Internal Defense, one of the Special Forces' missions.
FOB Forward Operating Base, a Special Forces tactical headquarters normally battalion size.
FORSCOM U.S. Army Forces Command.
FSG Family Support Group.
FTX Field Training Exercise.
EUCOM European Command.
G1 The Army's designation for a personnel officer or section for Division and above units.
G2 The Army's designation for an intelligence officer or section for Division and above units
G3 The Army's designation for an operations/plans officer or section for Division and above units.
G4 The Army's designation for a supply/logistics officer or section for Division and above units.
GEN General or O10.
Gs Guerrilla or indigenous forces trained by Special Forces to overthrow a country's government.
HALO High Altitude Low Opening, a speciality within Special Forces.
Helipad Helicopter landing pad.
HHC Headquarters and Headquarters Company.
ID Infantry Division.
IG Inspector General.
ITV Improved TOW Vehicle.
J1 The Armed Services designation for a personnel offer or section for a command consisting of members from different armed services like Air Force, Navy, Army and/or Marines.
J2 The Armed Services designation for an intelligence officer or section for a command consisting of members from different armed services like Air Force, Navy, Army and/or Marines.
J3 The Armed Services designation for an operations/plans officer or section for a command. consisting of members from different armed services like Air Force, Navy, Army and/or Marines.
J4 The Armed Services designation for a supply/logistics officer or section for a command consisting of members from different armed services like Air Force, Navy, Army and/or Marines.
JSOTF Joint Special Operations Task Force.
Kaserne German word *Kaserne* (plural: *Kasernen*), "barracks." It is the typical term used when naming the garrison location for NATO forces stationed in Germany.
LCM Landing Craft Mechanized; one of the Navy's a troop transport ships.
LLDB Luc Luong Dac Biet. South Vietnamese Army's Special Forces.
LTC Lieutenant colonel or O5.
LTG Lieutenant general or 09.
LZ Landing Zone.
MACO Marshaling Area Control Officer.
MACV U.S. Military Assistance Command, Vietnam.
MAJ Major or 04.
MCC Military Coordination Center.
MEDEVAC Medical Evacuation normally by helicopter.
MG Major general or 08.
MILPERCEN Military Personnel Center located in Washington D.C.
MOS Military Occupational Specially.
MSG Master Sergeant E8.
MTT Mobil Training Team.
NCO Noncommissioned Officer.
NGO Non-governmental Organizations.
NVA North Vietnamese Army.
O&I Operations and Intelligence.
OCS Officers' Candidate School.
ODA Operational Detachment A.
OER Officer Evaluation Report.

OER Officer Efficiency Report.
OIC Officer in Charge.
OPC II Operation Provide Comfort II.
OPCOM Operational Command.
OPORD Operations Order.
ORB Officer Records Brief.
ORT Operational Readiness Testing.
OTP One-time pads; crypto pads about the size of a small notebook.
PCS Permanent Change of Station.
PIRs Principal Indigenous Rations, the green tinfoil bags of dried food that we provided the CIDG for field operations.
PL Patrol Leader.
PLF Parachute landing fall.
PRP Personal Readiness Program.
PSP Perforated Steel Planking; strips of steel sapped together to form a landing stip or helicopter landing pad.
PT Physical training.
Q Course Special Forces Qualification Course.
RAF Royal Air Force, British air forces.
RG Readiness Group.
RI Ranger Instructor.
RIF Reduction in Force.
RON Rest Over Night.
S1 The Army's designation for a personnel officer or section for Brigade and lower units.
S2 The Army's designation for an intelligence officer or section for Brigade and lower units.
S3 The Army's designation for a operations/plans officer or section for Brigade and lower units.
S4 The Army's designation for a supply/logistics officer or section for Brigade and lower units.
SADM Small Atomic Demolitions Munitions.
SCIF Sensitive Compartmented Information Facility.
SCUBA Team A team that specialized in underwater operations.

SERE Survival, Evasion, Resistance, and Escape.
SF Special Forces; Green Berets.
SFC Sergeant First Class E7.
SFOB Special Forces Operational Base, tactical unit normally a Special Forces Group level unit.
SFQC Special Forces Qualification Course.
SLDC Shitty Little Detail Colonel.
SGM Sergeant Major E9.
SGT Buck sergeant or E5.
SITREP Situation Report.
SOCEUR Special Operations Command Europe.
SOCK Special Operations Command Korea.
SOCOM Special Operations Command.
SOP Standard Operating Procedure.
SSG Staff Sergeant E6.
STANO Surveillance Target Acquisition and Night Observation.
SWC U.S. Army John F. Kennedy Special Warfare Center and School.
TA50 Equipment issued to soldiers that they take to the field like rucksacks, sleeping bags, ponchos and such.
TDY Temporary Duty; short one-week to three-weeks away from the headquarters, often with extra pay.
TOC Tactical Operations Center.
TOWs Heavy antitank weapon.
TPIs Technical Proficiency Inspections.
UDT Vest Life jacket/vest worn when diving, and inflated when needed by activating a CO2 cartage.
USAJFKSWCS U.S. Army John F. Kennedy Special Warfare Center and School–known informally as SWC.
USAREUR U.S. Army Europe.
USASOC United States Army Special Operations Command.
USASFC United States Army Special Forces Command.

USFK Headquarters, U.S. Forces Korea.
USPA United States Parachute Association.
USSOCOM United States Special Operations Command.
UW Unconventional Warfare, one of the Special Forces missions.
UWO Underwater Operations or SCUBA; a speciality within Special Forces.
VC Viet Cong, an arm of the North Vietnamese fighting in South Vietnam.
VI Voluntary Indefinite.
WMD Weapons of Mass Destruction.
WO Warrant Officer.
WSI Water Safety Instructor.
XO Executive Officer.

Prologue: Stumbling Toward Enlightenment

A Memoir by Polly B. Davis

Blindly I bumbled off to college, any plans for the future eluding me. In the recesses of my mind, I would graduate, marry the ideal man, rear exceptional children, a boy and a girl of course, and be an extraordinary wife and mother. Naturally, this family would be an asset to the state of Georgia.

At the University during orientation, I was one of the 1,000 freshmen girls who were herded into a whitewashed auditorium reeking of body heat and tension; not one building on campus was air conditioned in 1963. On the ninth floor of my nine-story dormitory, we already knew humidity and underarm perspiration first hand.

My eyes flickered over the monoculture crowd. A colony of nuns offered more diversity than this flock of young girls, all feathered in A-line skirts, Peter Pan collared blouses, and weejuns. The dean of students, jacket tucked under his elbow, shirt glued to his back, took the steps up to the stage sideways, eyeing the plethora of pretty girls with eyes on him. The late summer heat had gotten to him, too, I supposed. He edged to the microphone and announced the plan in staccato: "If you want be a nurse, go to rooms 151 and 152; if you want to teach, stay here; if you plan to do anything else, go to room 103," a large closet of a room, I'd noticed, when I passed by it earlier and wondered about the sign taped to the door: All Other Majors. I remained seated, curiously observing those exiting, watching for girls I knew from Macon, my hometown. Only one left. There in the "education section," we remained true to our mothers' advice: "Get a teaching degree, just in case" [or] "to fall back on," meaning if your husband died or left you and the kids.

My roommate from South Georgia fixed me up with Tom. I was sure we were a perfect match: both Baptists, both blond and blue eyed, both raw oyster crazy, and our people were from South Georgia. In my naiveté he met all the criteria. We ended up traipsing the world together, often at odds, gradually realizing the path we followed could easily spell disaster for a marriage.

Vietnam loomed, but students at the University of Georgia were too absorbed in sorority and frat houses, keg parties, and rush to be concerned with a war in that tiny little country. We lived in a vacuum, a microcosm of polite society, libraries, and naïve kids. Recruiters would surely not invade a southern college campus like the University of Georgia.

Graduation was approaching, and the family friend running the draft board in Vienna, Georgia, Tom's hometown, called to alert him that his number was coming up. A call to his dad, whom he thought might pull some strings, was to no avail: "Son, I'd hate to see you go to war, but be a good patriot. In this case, I want you to be real careful and write home a lot." The immediacy was buffered only by the privilege of beginning law school the summer following graduation. After that short stint, he joined the Army. As it turned out, Tom found that following in his grandfather and dad's footsteps would not be his legacy.

When we married after basic training and Officer Candidate School, I stood clueless at the altar. The only soldiers I'd known were in the comics. The profession was one I didn't hold in high regard, and marrying one seemed surreal.

Our first tour at Fort Bragg in Fayetteville, North Carolina, focused on preparation for the war. Fighting the fear that filled our throats when TV's newscasters flashed body bags and rising numbers, we were forced to pretend life was normal. Finding a place to live until we moved onto Post foreshadowed the rest of our life in the army. Tom had heard of some apartments that were really cheap. He always went for "cheap," and his decision was made. Convincing me was easy: his income as a second lieutenant was meager. I had no job, no income, and applying for a job teaching was pointless: public school had already begun.

Regardless, Tom insisted we check at the school board office. There they sent me to meet the burly principal of a rural high school. Stubbing his cigar out in the can outside his office, he led me in and pumped my hand heartily as though we'd already signed a pact. Unconcerned that I'd majored in English, he revealed that he did have a teaching position available in the social sciences, adding, "You know, you can teach just about anything you want in a sociology class." After a short discussion, he cocked his head, leaned into my eyes and finished with, "Let's make a deal: I need a teacher, and you need a job."

This brand new high school, spread like a college campus, sat on several acres out in the far reaches of the county. The first of its kind in Cumberland County, it was a melting pot in almost even percentages of Caucasian, Black, and Native American students. I had never attended a school with mixed races, and this confluence of students, so ignorant of, yet so intrigued with the others' cultures, filled Cape Fear High School with a positive energy such as I'd never known. Between teaching a subject slightly foreign to me and working with students so hungry for knowledge, I received an education I'd never have expected. The setting proved perfect for an interactive sociology class, which even strayed into methods of birth control. My favorite co-teacher and I shared a planning period each day. When I humorlessly mentioned the strong odor emitted by a student who chose the desk on the front row, she told me, gently, "I'm pretty sure he's never lived in a house with a bath tub or a shower."

For the first month or so of this tour, Tom and I shared an automobile. We left for Ft. Bragg in absolute darkness with a hint of the moon frowning down on us. I deposited him at a one-story wooden structure left over from WWII days, turned around, and continued the twenty miles east to work, watching daylight eek its way over the line of pines. In the afternoon, the trip was reversed.

Once Tom was jump qualified, his enthusiasm for parachuting peaked. He joined the 82nd Sports Parachute Club, and we routinely spent weekends at the drop-zone. I was the "good wife" who grudgingly prepared the picnic lunch and sat around with other wives waiting for our men to jump, cheering them on, and waiting, waiting, waiting until the jumps were over and we would continue on to the Club for beers with the guys and war stories.

This grew old. Since I believed, in those days, that husbands and wives should do everything together, I decided to join him. After investigating, I found that, yes, spouses and civilians, anybody for that matter, could take sky diving lessons on Post. I joined the Club.

For four weeks, I drove onto post in the evenings for jump classes, wondering to myself exactly what I was trying to prove. We met the instructor in the riggers' shed, a dark ponytail held back by a rubber band the only thing distinguishing him from the male students. A civilian master sports parachutist, over six-feet, with a strong profile and straight teeth, he smiled often and treated us equally, although he was slightly taken aback to find a women in his class. I breathed a sign of relief that he treated me no differently that the men. We learned to pack our own parachutes, the ones we would be jumping, and to understand the elements and nuances of jumping out of a plane. We also learned how to make a safe parachute landing fall, rolling in a sawdust pile for the better part of each evening. Back home I had to shake the sawdust out of my hair, my boots, and my clothes, down to my underwear.

The day arrived when I was qualified to make my first static line jump. The forty pounds of olive drab gear cinched tightly to my back hobbled my movement. I found it difficult to breathe, and my vision was impaired by the protective helmet. Did I still want to do this? The instructor continued to encourage me as we boarded the small Cessna 170 together. When the pilot cut the motor, I was frozen between nausea and fear of blacking out. The motor purred quietly, but wind whipped in the open door. I was instructed, "Reach out and grab the strut with both hands, stabilize yourself, then let go."

So I reached out the door. Mind numb, fist in my chest, I was able to clutch the strut, but the thundering gusts outside the plane quickly blew me into a split second of chaos, a terrible spinning and sputtering, and then I heard it: a smooth popping sound, a snap, a rough jerk, and my chute floated open above me. I was alive, and all was calm. No sound, no sense of movement, aquamarine heavens, and billowing groves of green below me, I floated, savoring the absolute ecstasy of the moment.

Suddenly voices below were shouting, "Use your toggle lines." Too late. I was slowly drifting into a grove of pine trees. Remembering the instructions to streamline my body if this happened, I slipped down between branches and was caught just a few feet off the ground. Rapidly jogging toward me, a young soldier was muttering loudly, "A woman just ought 'a stay home; she just ought 'a stay home where she belongs."

The next jump boded better. Out the door spinning, but keeping my wits about me, I pulled the toggle lines to guide my open chute toward the waiting crowd. Dumb luck or determination propelled me to a landing fifteen feet from the center. Cheers all around. My decision was not difficult to make: "Yep, I'm quitting while I'm ahead." Tom admitted, "I haven't had a good night's sleep since you started this." Years passed, and we knew many a paratrooper who lost his life. To this day, I believe it's suicidal for people to jump out of planes for sport, especially those with families depending on them.

Shortly after I began teaching, the commander's wife, or "Commandant" as I later heard other wives whisper behind her back, called to inform me of a baby shower for another wife, and I should bring cookies and a gift. When I explained that I was working, she hruuumped and told me, "That's highly unusual, and I hope you'll find a way to be there regardless." I hung up and complained: "A lady I don't know just called to invite, rather order, me to a shower for someone I don't know and told me what to bring for refreshments and a gift. She had some nerve." Tom was baffled also. But when I told him who had called, he looked concerned. We both had a lot to learn. It didn't take long to realize why wives banded together as a support system: it replaced the family and friends that we all had left behind. A southern commander's wife, I thought, would have handled this more tactfully.

Vietnam loomed like a forest fire edging past the break. Tom went to war, and I went to graduate school. No longer the crazy coed, I saw Athens as the stay for young men who piled degrees upon degrees to stay in college. I'd catch myself avoiding windows, the look-out for those gray cars that pulled up in front of homes, shadowy figures forcing their feet to move to doorways with that awful message. We were lucky.

Tom suggested that I take a scuba class while in Athens so we could dive together on our next tour. The week he returned from Vietnam I graduated with my masters and flunked the scuba class the same day. Humiliating. Defending a thesis and learning those abominable decompression tables simultaneously had been hopeless. Later, when we moved to Fort Devens, twenty minutes from Concord, Massachusetts, Tom and I tackled another scuba class together. We passed without a hitch and completed our check-out dive in Walden Pond. That, for an English major, was pure serendipity.

Still, we were in the military, and time was nagging us to make the decision to stay in or to get out. The trick was on me: my husband was living his dream. He was now employed in the daring, sporting club of the Special Forces: snow skiing, scuba diving, parachuting, and mountain climbing. For him, the decision demanded little thought. For me, shaking up the system became my forte. As Tom's "dependant" wife, I challenged the term from the Bavarian Alps to the North Carolina Sandhills in my quest for one wife's independence.